D1085798

THE GREAT MATHEMATICIANS

THE GREAT MATHEMATICIANS

UNRAVELLING THE MYSTERIES OF THE UNIVERSE

Raymond Flood and Robin Wilson

Raymond Flood is an emeritus fellow and former vice president of Kellogg College, Oxford and was previously university lecturer in computing studies and mathematics at the Department for Continuing Education, Oxford University. His main research interests lie in statistics and the history of mathematics, and he was formerly president of the British Society for the History of Mathematics.

Robin Wilson is an emeritus professor of pure mathematics at the Open University, emeritus professor of geometry at Gresham College, London, and a former fellow of Keble College, Oxford. He currently teaches at Pembroke College, Oxford, and is president-elect of the British Society for the History of Mathematics. He is involved with the popularization and communication of mathematics and, in 2005, he was awarded a Pólya prize by the Mathematical Association of America for 'outstanding expository writing'.

This edition published in 2013 by Arcturus Publishing Limited
26/27 Bickels Yard, 151–153 Bermondsey Street,
London SE1 3HA

Copyright © 2011 Arcturus Publishing Limited

All rights reserved. No part of this publication may be reproduced, stored in a retrieval system, or transmitted, in any form or by any means, electronic, mechanical, photocopying, recording or otherwise, without prior written permission in accordance with the provisions of the Copyright Act 1956 (as amended). Any person or persons who do any unauthorised act in relation to this publication may be liable to criminal prosecution and civil claims for damages.

ISBN: 978-1-84837-902-2
AD001763EN

Printed in China

PREFACE

This book aims to present mathematics 'with a human face' and to celebrate its achievements in their historical context. It features a personal selection of many of the mathematicians whose lives and work interest us, presented with as little technical background as possible. Due to limitations of space and the double-page format, we have had to omit several mathematicians that we should have liked to include, and said less than we wished about those we have featured, but we hope that you will find the result interesting and that it will whet your appetite for further reading — we have included some suggestions for this at the end.

This is not a book on the history of mathematics, which is a vast subject demanding a very different treatment. Our featured mathematicians are organized roughly chronologically, with a mathematical timeline and some maps appearing at the beginning of the book. Throughout, we have tried to present the ideas and results in modern terminology and notation, so as to make them more accessible, and extracts have been translated into English rather than appearing in the original languages.

We wish to thank June Barrow-Green, Jacqueline Stedall, George Bitsakakis and Benjamin Wardhaugh for reading and commenting on portions of the manuscript, and also Nigel Matheson and his colleagues at Arcturus Publishing.

Finally, a book of this scope inevitably contains errors and omissions. Each author wishes to make it clear that these are entirely the fault of the other.

Raymond Flood and Robin Wilson

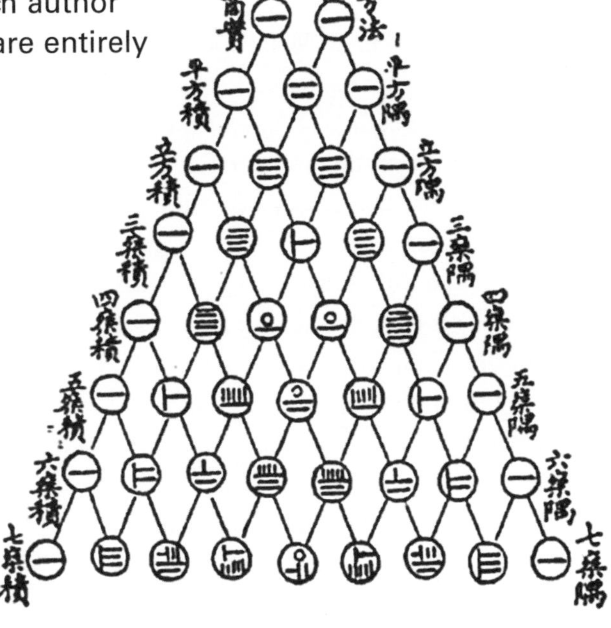

An arithmetical triangle (later known as Pascal's triangle) from Zhu Shijie's *Siyuan yujian* (Precious Mirror of the Four Elements) of 1303

CONTENTS

INTRODUCTION

The stories of Isaac Newton and the apple, and of Archimedes running naked along the street shouting 'Eureka', are familiar to many. But which mathematicians are the answers to the following questions?

- **Who was killed in a duel?**
- **Who published books, yet did not exist?**
- **Who was crowned Pope?**
- **Who were Dr Mirabilis and Dr Profundus?**
- **Who learned calculus from her nursery wallpaper?**
- **Who was excited by a taxicab number?**
- **Who measured the chests of 5732 Scottish soldiers?**

And what have Geoffrey Chaucer, Christopher Wren, Napoleon, Florence Nightingale and Lewis Carroll to do with mathematics?

As these questions may indicate, and as the pages of this book will show, mathematics has always been a human endeavour as people have found themselves grappling with a wide range of problems, both practical and theoretical. The subject has as long and interesting a history as literature, music or painting, and its origins were both international and multicultural.

For many who remember mathematics from their schooldays as a dull and dusty subject, largely incomprehensible and irrelevant to their everyday lives, this view of mathematics may come as a surprise. The subject has all too often been presented as a collection of rules to be learned and techniques to be applied, providing little understanding of the underlying principles or any appreciation of the nature of the subject as a whole — it is rather like teaching musical scales and intervals without ever playing a piece of music.

For wherever we look, mathematics pervades our daily lives. Our credit cards and the nation's defence secrets are kept secure by encryption methods based on the properties of prime numbers, and mathematics is intimately involved when one flies in a plane, starts a car, switches on the television, forecasts the weather, books a holiday on the internet, programs a computer, navigates heavy traffic, analyses a pile of statistical data, or seeks a cure for a disease. Without mathematics as its foundation there would be no science.

Mathematicians are often described as 'pattern-searchers' — whether they study abstract patterns in numbers and shapes or look for symmetry in the natural world around us. Mathematical laws shape the patterns of seeds in sunflower heads and guide the solar system that we live in. Mathematics analyses the minuscule structure of the atom and the massive extent of the universe.

But it can also be a great deal of fun. The logical thinking and problem-solving techniques that one learns in school can equally be put to recreational use. Chess is essentially a mathematical game, many people enjoy solving logical puzzles based on mathematical ideas, and thousands travel into work each day struggling with their sudoku puzzles, a pastime arising from combinatorial mathematics.

Mathematics is developing at an ever-increasing rate — indeed, more new

William Blake's 'Isaac Newton'

mathematics has been discovered since the Second World War than was known up to that time. An outcome of all this activity has been the International Congresses of Mathematicians that are held every four years for the presentation and discussion of the latest advances.

But none of this would have happened if it had not been for the mathematicians who created their subject.

In this book you will meet time-measurers like the Mayans and Huygens, logicians like Aristotle and Russell, astronomers like Ptolemy and Halley, textbook writers like Euclid and Bourbaki, geometers like Apollonius and Lobachevsky, statisticians like Bernoulli and Nightingale, architects like Brunelleschi and Wren, teachers like Hypatia and Dodgson, arithmeticians like Pythagoras and al-Khwarizmi, number-theorists like Fermat and Ramanujan, applied mathematicians like Poisson and Maxwell, algebraists like Viète and Galois, and calculators like Napier and Babbage. We hope that you find all their lives and achievements as fascinating as we do.

MAPS

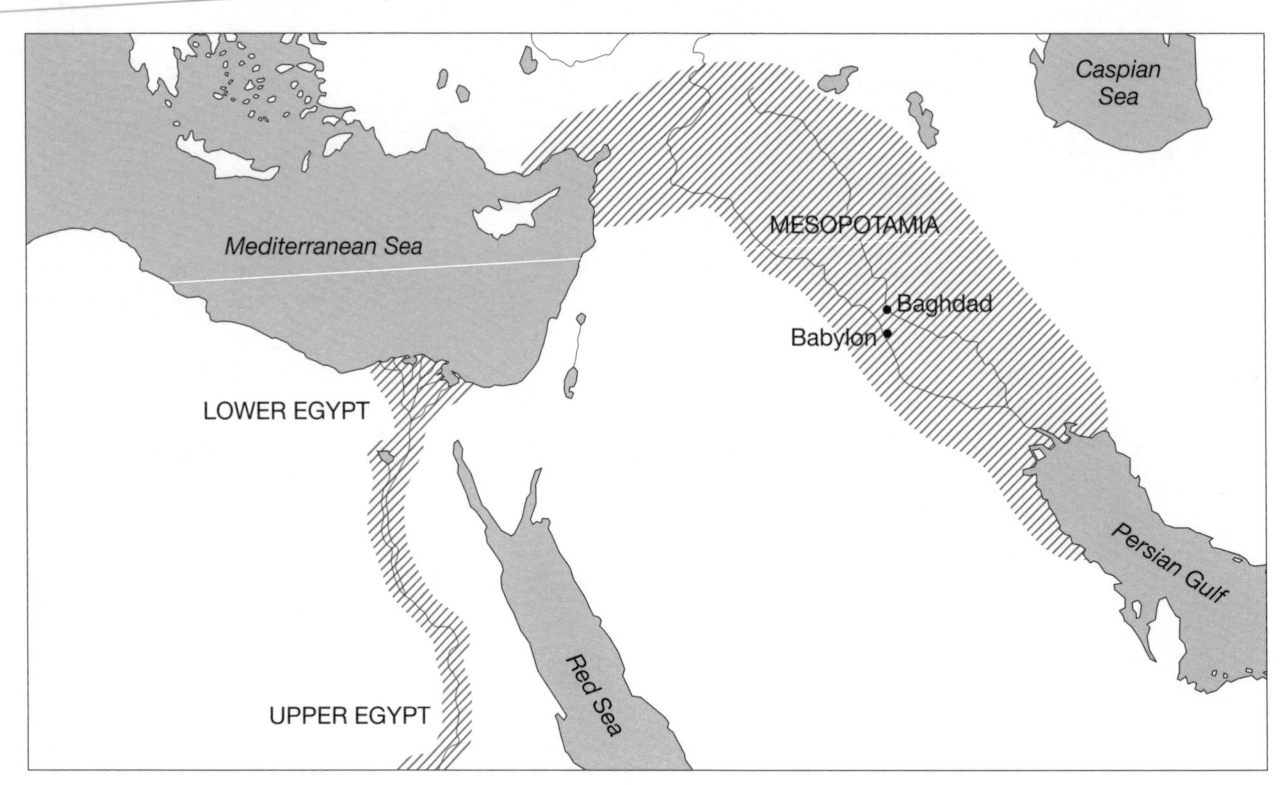

Egypt and Mesopotamia

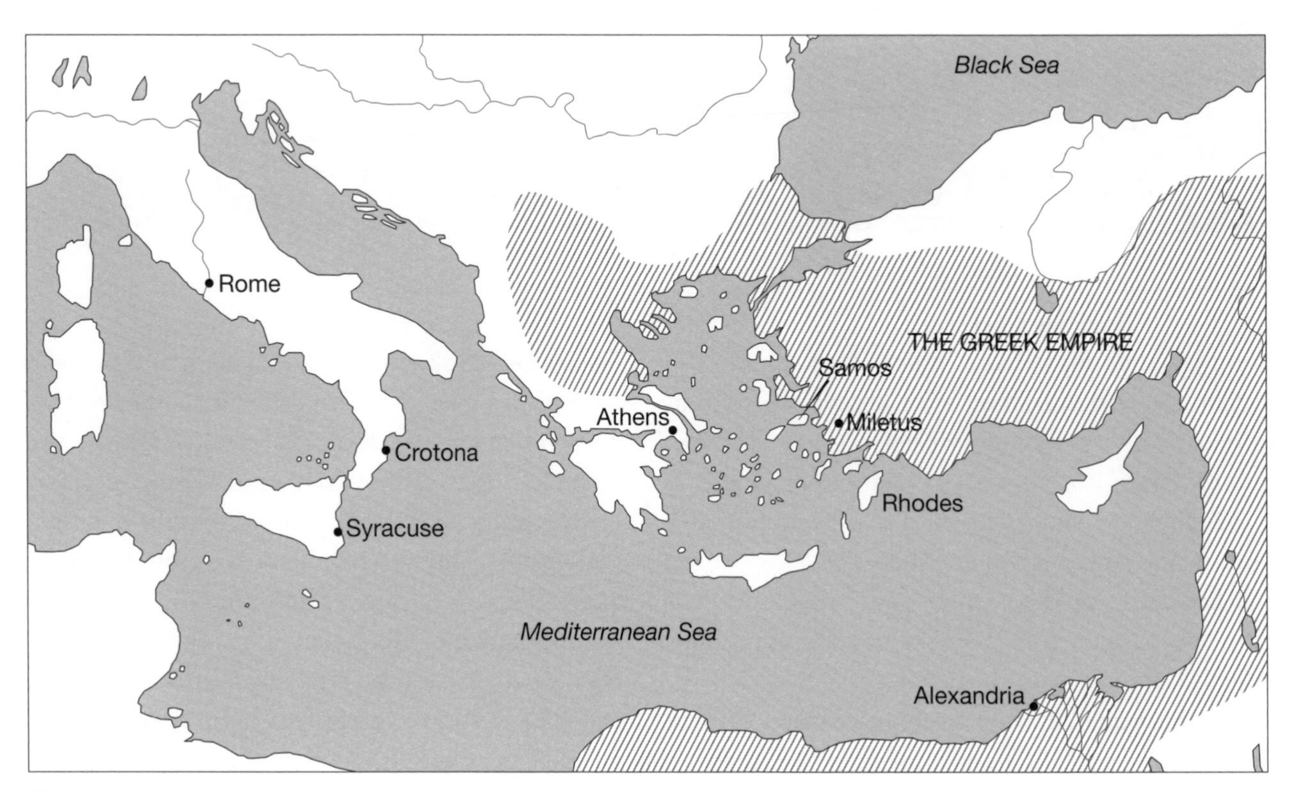

Greece

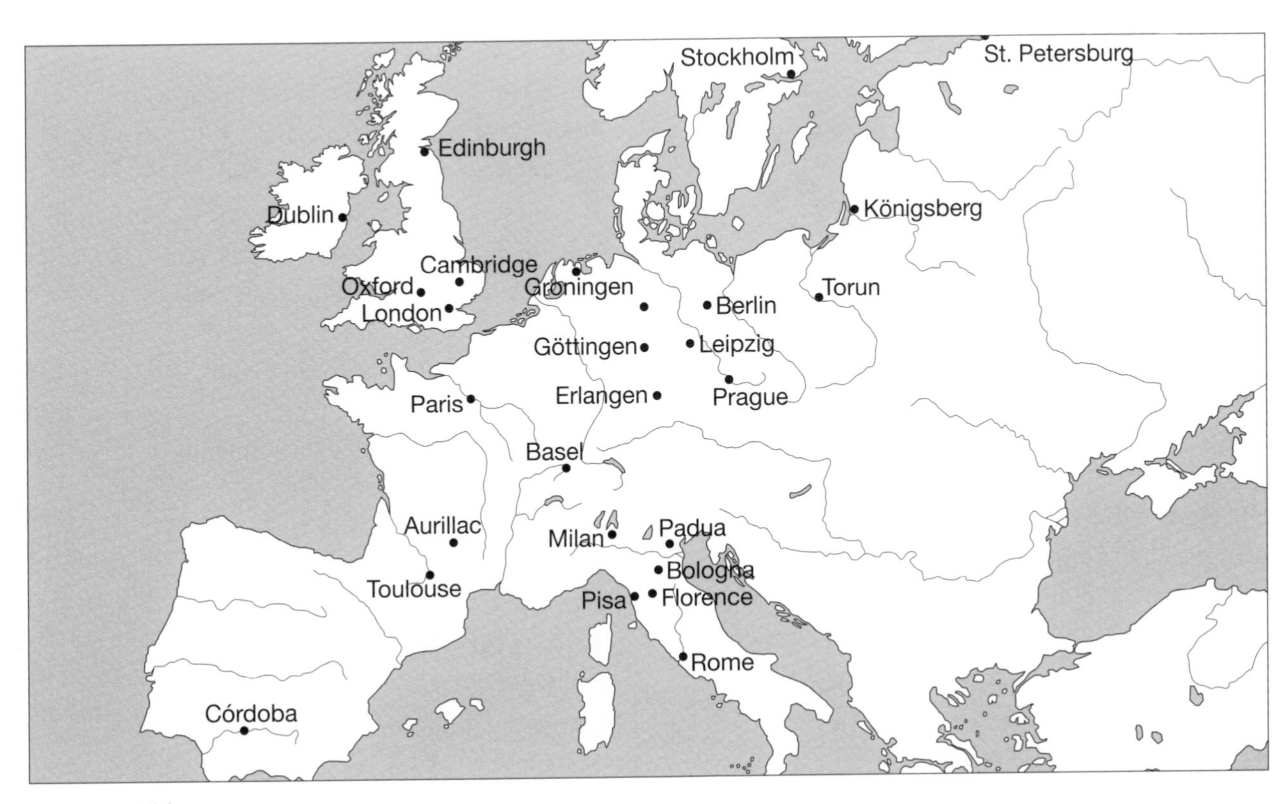

European cities

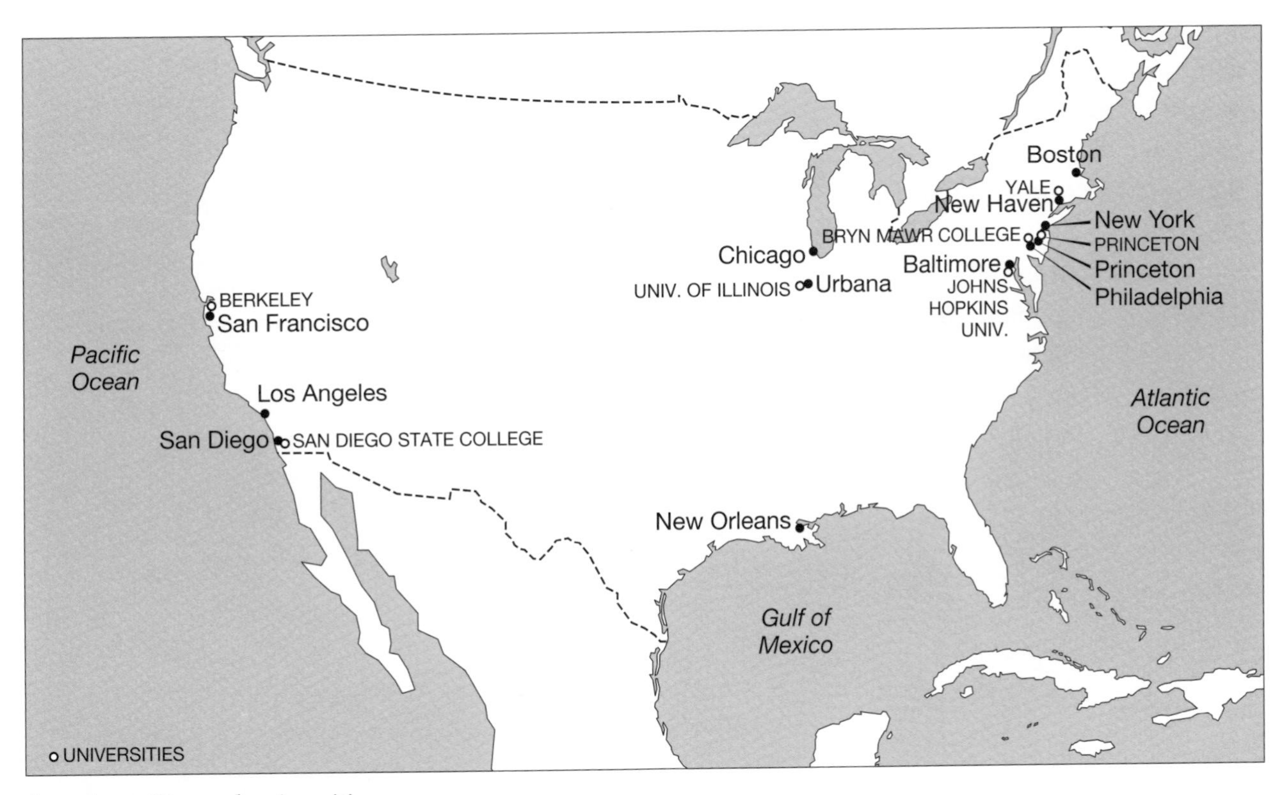

American cities and universities

TIMELINE

BC	
c.1850	Moscow papyrus, Egypt
c.1800	Old Babylonian mathematics
c.1650	Rhind papyrus, Egypt
c.624–c.546	Thales
c.570–490	Pythagoras
429–347	Plato
384–322	Aristotle
c.300	Euclid
c.287–212	Archimedes
c.262–c.190	Apollonius
c.200?	*Jiu zhang suan shu*
190–120	Hipparchus
c.100?	*Zhou bi suan jing*

AD	
c.60–120	Nicomachus of Gerasa
c.100–c.170	Ptolemy of Alexandria
c.250	Diophantus
c.290–c.350	Pappus
c.360–415	Hypatia
b.476	Aryabhata the Elder
c.480–524	Boethius
598–670	Brahmagupta
c.783–c.850	al-Khwarizmi
c.940–1003	Gerbert of Aurillac
965 –1039	Alhazen (ibn al-Haitham)
1048 –1131	Omar Khayyam
c.1170–1240	Leonardo of Pisa (Fibonacci)
c.1175–1253	Bishop Grossteste
c.1200	Mayan Dresden codex
c.1214–1294	Roger Bacon
c.1290–1349	Thomas Bradwardine
1292–1336	Richard of Wallingford
c.1323–1382	Nicole Oresme
1342–1400	Geoffrey Chaucer
1377–1446	Filipo Brunelleschi
1404–1472	Leon Battista Alberti
c.1415–1492	Piero della Francesca
1447–1517	Luca Pacioli
1452–1519	Leonardo da Vinci
1471–1528	Albrecht Dürer
1473–1543	Nicolaus Copernicus
c.1500–1557	Niccolò of Brescia (Tartaglia)
1501–1576	Gerolamo Cardano
1510–1558	Robert Recorde
1512–1594	Gerardus Mercator
c.1526–1572	Rafael Bombelli
1540–1603	François Viète
1550–1617	John Napier
1560–1621	Thomas Harriot
1561–1630	Henry Briggs
1564–1642	Galileo Galilei
1571–1630	Johannes Kepler
1588–1648	Marin Mersenne
1591–1661	Girard Desargues
1596–1650	René Descartes
1598–1647	Bonaventura Cavalieri
1601–1665	Pierre de Fermat
1601–1680	Athanasius Kircher
1602–1675	Gilles Personne de Roberval
1616–1703	John Wallis
1623–1662	Blaise Pascal
1629–1695	Christiaan Huygens

1632–1723 Christopher Wren
1635–1703 Robert Hooke
1642–1727 Isaac Newton
1646–1716 Gottfried Leibniz
1654–1705 Jacob Bernoulli
1656–1742 Edmond Halley
1667–1748 Johann Bernoulli
1706–1749 Emilie du Châtelet
1707–1783 Leonhard Euler
1717–1783 Jean le Rond d'Alembert
1736–1813 Joseph-Louis Lagrange
1746–1818 Gaspard Monge
1749–1827 Pierre-Simon Laplace
1768–1830 Joseph Fourier
1776–1831 Sophie Germain
1777–1855 Carl Friedrich Gauss
1781–1840 Siméon Denis Poisson
1788–1867 Jean Victor Poncelet
1789–1857 Augustin-Louis Cauchy
1790–1868 August Möbius
1791–1871 Charles Babbage
1792–1856 Nikolai Lobachevsky
1793–1841 George Green
1802–1829 Niels Henrik Abel
1802–1860 János Bolyai
1816–1852 Ada, Countess of Lovelace
1805–1865 William Rowan Hamilton
1806–1895 Thomas Penyngton Kirkman
1811–1832 Évariste Galois
1814–1897 James Joseph Sylvester
1815–1864 George Boole
1819–1903 George Gabriel Stokes
1820–1910 Florence Nightingale
1821–1894 Pafnuty Chebyshev
1821–1895 Arthur Cayley
1824–1907 William Thomson (Lord Kelvin)
1826–1866 Bernhard Riemann
1831–1879 James Clerk Maxwell
1831–1901 Peter Guthrie Tait
1832–1898 Charles Dodgson
1845–1918 Georg Cantor
1849–1925 Felix Klein
1850–1891 Sonya Kovalevskaya
1854–1912 Henri Poincaré
1862–1943 David Hilbert
1864–1909 Hermann Minkowski
1872–1970 Bertrand Russell
1877–1947 Godfrey Harold Hardy
1879–1955 Albert Einstein
1882–1935 Emmy Noether
1885–1977 John Edensor Littlewood
1887–1920 Srinivasa Ramanujan
1903–1957 John von Neumann
1906–1978 Kurt Gödel
1912–1954 Alan Turing
1919–1985 Julia Robinson
1924–2010 Benoit Mandelbrot
b.1928 Wolfgang Haken
b.1932 Kenneth Appel
b.1934 Nicolas Bourbaki
1936 First Fields Medallists
b.1947 Yuri Matiyasevich
b.1953 Andrew Wiles
b.1966 Grigori Perelman

CHAPTER 1

ANCIENT MATHEMATICS

Mathematics is ancient and multicultural. Several examples of early counting devices on bone (such as tally sticks) have survived, and some of the earliest examples of writing (from around 5000BC) were financial accounts involving numbers. Much mathematical thought and ingenuity also went into the construction of such edifices as the Great Pyramids, the stone circles of Stonehenge, and the Parthenon in Athens.

In this chapter we describe the mathematical contributions of several ancient cultures: Egypt, Mesopotamia, Greece, China, India and Central America. The mathematics developed in each culture depended on need, which may have been practically inspired (for example, agricultural, administrative, financial or military), academically motivated (educational or philosophical), or a mixture of both.

SOURCE MATERIAL

Much of what we know about a culture depends on the availability of appropriate primary source material.

For the Mesopotamians we have many thousands of mathematical clay tablets that provide much useful information. On the other hand, the Egyptians and the Greeks wrote on papyrus, made from reeds that rarely survive the ravages of the centuries, although we do have two substantial Egyptian mathematical papyri and a handful of Greek extracts. The Chinese wrote their mathematics on bamboo and paper, little of which has survived. The Mayans wrote on stone pillars called *stelae* that contain useful material. They also produced codices, made of bark paper; a handful of these survive, but most were destroyed during the Spanish Conquest many centuries later.

Apart from this, we have to rely on commentaries and translations. For the classical Greek writings we have

A Mesopotamian clay tablet

commentaries by a few later Greek mathematicians, and also a substantial number of Arabic translations and commentaries by Islamic scholars. There are also later translations into Latin, though how true these may be to the original works remains a cause for speculation.

COUNTING SYSTEMS

All civilizations needed to be able to count, whether for simple household purposes or for more substantial activities such as the construction of buildings or the planting of fields.

As we shall see, the number systems developed by different cultures varied considerably. The Egyptians used a decimal system with different symbols for 1, 10, 100, 1000, etc. The Greeks used different Greek letters for the units from 1 to 9, the tens from 10 to 90, and the hundreds from 100 to 900. Other cultures developed place-value counting systems with a limited number of symbols: here the same symbol may play different roles, such as the two 3s in 3835 (referring to 3000 and 30). The Chinese used a decimal place-value system, while the Mesopotamians had a system based on 60 and the Mayans developed a system mainly based on 20.

Any place-value system needs the concept of zero; for example, we write 207, with a zero in the tens place, to distinguish it from 27. Sometimes the positioning of a zero was clear from the context. At other times a gap was left, as in the Chinese counting boards, or a zero symbol was specifically designed, as in the Mayan system.

The use of zero in a decimal place-value system eventually emerged in India and elsewhere, and rules were given for calculating with it. The Indian counting system was later developed by Islamic mathematicians and gave rise to what we now call the *Hindu–Arabic numerals,* the system that we use today.

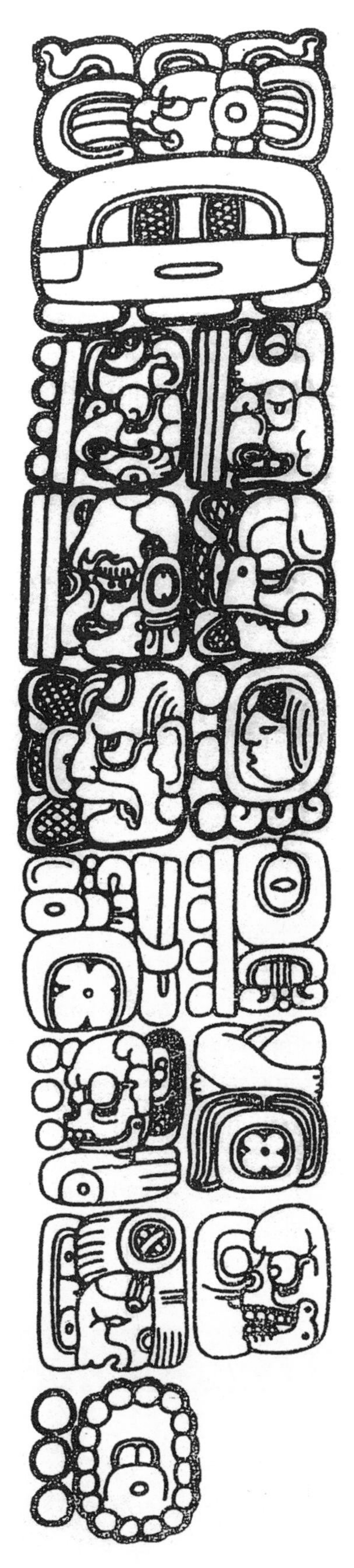

A Central American stela featuring Mayan head-form numbers

So, starting from the natural numbers, 1, 2, 3, ... , generations of mathematicians obtained all the *integers* — the positive and negative whole numbers and zero. This was a lengthy process that took thousands of years to accomplish.

THE EGYPTIANS

The magnificent pyramids of Giza, dating from about 2600BC, attest to the Egyptians' extremely accurate measuring ability. In particular, the Great Pyramid of Cheops, constructed from over two million blocks averaging around two tonnes in weight, is an impressive 140 metres high and has a square base whose sides of length 230 metres agree to within less than 0.01%.

Our knowledge of later Egyptian mathematics is scanty, and comes mainly from two primary sources: the 5-metre-long 'Rhind papyrus' (c.1650BC), named after its Victorian purchaser Henry Rhind and housed in the British Museum, and the 'Moscow papyrus' (c.1850BC), currently housed in a Moscow museum.

These papyri include tables of fractions and several dozen solved problems in arithmetic and geometry. Such exercises were used in the training of scribes, and range from division problems involving the sharing of loaves in specified proportions to those requiring the volume of a cylindrical granary of given diameter and height.

The pyramids of Giza

THE EGYPTIAN COUNTING SYSTEM

The Egyptians used a decimal system, but wrote different symbols (called *hieroglyphs*) for 1 (a vertical rod), 10 (a heel bone), 100 (a coiled rope), 1000 (a lotus flower), etc.

Each number appeared with the appropriate repetitions of each symbol, written from right to left; for example, the number 2658 was

The Egyptians calculated with *unit fractions* (or reciprocals), fractions with 1 in the numerator such as $^1/_8$, $^1/_{52}$ or $^1/_{104}$ (they also used the fraction $^2/_3$); for example, they wrote $^1/_8$ $^1/_{52}$ $^1/_{104}$ instead of $^2/_{13}$, since $^1/_8 + ^1/_{52} + ^1/_{104} = ^2/_{13}$.

To aid such calculations the Rhind papyrus includes a table of unit fractions for each of the fractions $^2/_5, ^2/_7, ^2/_9, \ldots, ^2/_{101}$.

The Egyptians' remarkable ability to calculate with these unit fractions can be seen in Problem 31 of the Rhind papyrus:

> *A quantity, its* $^2/_3$, *its* $^1/_2$ *and its* $^1/_7$, *added together become* 33. *What is the quantity?*

To solve this problem with our modern algebraic notation, we would call the unknown quantity x and obtain the equation

$$x + ^2/_3x + ^1/_2x + ^1/_7x = 33.$$

We would then solve this

THE AREA OF A CIRCLE

Several problems in the Rhind papyrus involve circles of a given diameter. You may recall that

The area of a circle of radius r is πr^2.

Since the diameter $d = 2r$, this area can also be written as $\frac{1}{4}\pi d^2$.

The number that we now denote by π also appears in the formula for the circumference:

The circumference of a circle of radius r and diameter d is $2\pi r = \pi d$.

The value of π is about $^{22}/_7$ (= $3^1/_7$), and a more accurate approximation is 3.1415926; however, π cannot be written down exactly as its decimal expansion goes on for ever.

Problem 50 of the Rhind papyrus asks for the area of a circle of diameter 9:

Example of a round field of diameter 9 khet. What is its area?

Answer: Take away $^1/_9$ *of the diameter, namely 1; the remainder is 8. Multiply 8 times 8; it makes 64. Therefore it contains 64 setat of land.*

The Egyptians found by experience that they could approximate the area of a circle with diameter *d* by reducing *d* by one-ninth and squaring the result. So here, where $d = 9$, they reduced *d* by one-ninth (giving 8) and then squared the result (giving 64).

Their method corresponds to a value of π of $3^{13}/_{81}$, which is about 3.16, within 1 per cent of the correct value.

equation to give $x = 14^{28}/_{97}$. But the answer the Egyptians gave, expressed with unit fractions, was

$14\ ^1/_4\ ^1/_{56}\ ^1/_{97}\ ^1/_{194}\ ^1/_{388}\ ^1/_{679}\ ^1/_{776}$

— a truly impressive feat of calculation.

DISTRIBUTION PROBLEMS

Several problems on the Rhind papyrus involve the distribution of some commodity, such as bread or beer. For example, Problem 65 asks:

Example of dividing 100 *loaves among* 10 *men, including a boatman, a foreman and a doorkeeper, who receive double shares. What is the share of each?*

To solve this, the scribe replaced each man receiving a double share by two people:

The working out. Add to the number of men 3 *for those with double portions; it makes* 13. *Multiply* 13 *so as to get* 100; *the result is* $7\ ^2/_3\ ^1/_{39}$. *This then is the ration for seven of the men, the boatman, the foreman, and the doorkeeper receiving double shares* [= $15^1/_3\ ^1/_{26}\ ^1/_{78}$].

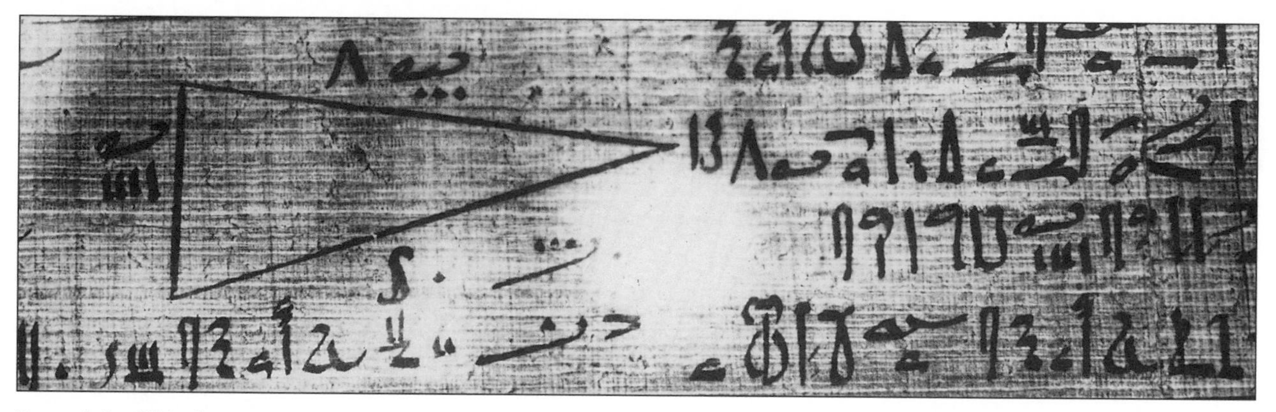

Part of the Rhind papyrus

THE MESOPOTAMIANS

Mesopotamian (or Babylonian) mathematics developed over some 3000 years and over a wide region, but the problems we consider here date mainly from the Old Babylonian period (around 1800BC). The word *Mesopotamian* comes from the Greek for 'between the rivers' and refers to the area between the rivers Tigris and Euphrates in modern-day Iraq.

The primary source material is very different in form and content from that of the Egyptians of the same period. Using a wedge-shaped stylus, the Mesopotamians imprinted their symbols into moist clay – this is called *cuneiform writing* – and the tablet was then left to harden in the sun. Many thousands of mathematical clay tablets have survived.

THE SEXAGESIMAL SYSTEM

We write numbers in the decimal place-value system, based on 10, with separate columns for units, tens, hundreds, etc., as we move from right to left. Each place has value ten times the next; for example, 3235 means

$$(3 \times 1000) + (2 \times 100) + (3 \times 10) + (5 \times 1).$$

The Mesopotamians also used a place-value system, but it was a 'sexagesimal' system, based on 60: each place has value sixty times the next. It used two symbols, which we write here as Y for 1 and < for 10:

- for 32 they wrote <<<YY;
- for 870 they wrote <YYYY <<<,
  since $870 = 840 + 30 = (14 \times 60) + 30$;
- for 8492 they wrote YY <<Y <<<YY,
  since $8492 = (2 \times 60^2) + (21 \times 60) + 32$.

Remnants of their sexagesimal system survive in our measurements of time (60 seconds in a minute, 60 minutes in an hour) and of angles. The Mesopotamians developed the ability to calculate with large sexagesimal numbers, and used them to chart the cycles of the moon and construct a reliable calendar.

TYPES OF TABLET

There were essentially three types of mathematical clay tablet. Some of them list tables of numbers for use in calculations and are called *table tablets:* an example of a table tablet is the 9-times multiplication table below.

Other clay tablets, called *problem tablets,* contain posed and solved mathematical problems. A third type may be described as *rough work,* created by students while learning.

An example of a Mesopotamian problem is the following, on the weight of a stone: it appears on a clay tablet featuring twenty-three problems

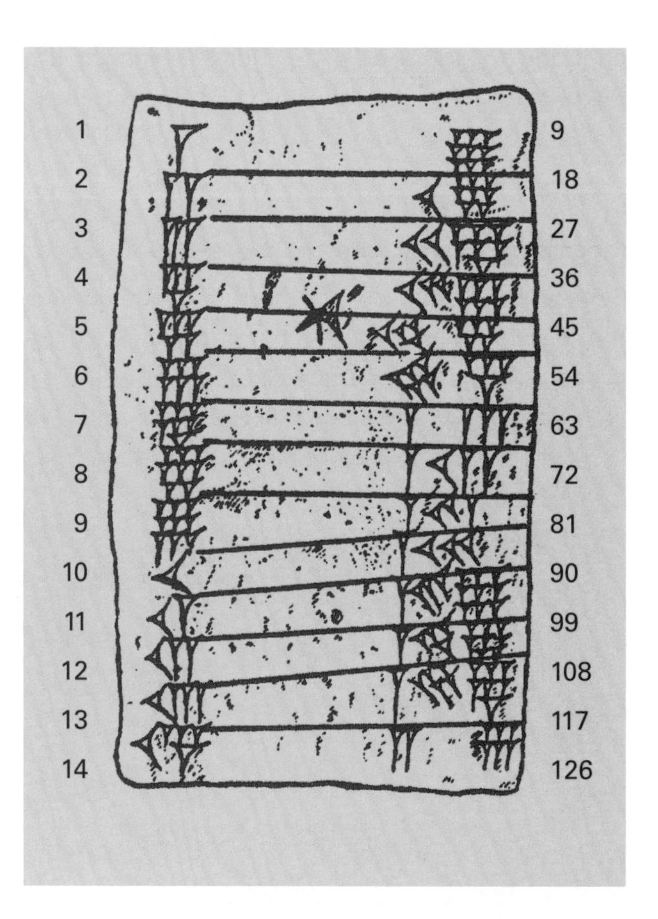

A drawing of a table tablet

THE SQUARE ROOT OF 2

A particularly unusual tablet, which illustrates the Mesopotamians' remarkable ability to calculate with great accuracy, depicts a square with its two diagonals and the sexagesimal numbers 30, 1;24,51,10 and 42;25,35. These numbers refer to the side of the square (of length 30), the square root of 2, and the diagonal (of length $30\sqrt{2}$).

The amazing accuracy of their value for the square root,

1;24,51,10

$= 1 + {}^{24}/_{60} + {}^{51}/_{3600} + {}^{10}/_{216000}$

(= 1.4142128... in decimal notation),

becomes apparent if we square it — we get

1;59,59,59,38,1,40

(= 1.999995... in decimal notation).

This differs from 2 by about 5 parts in a million.

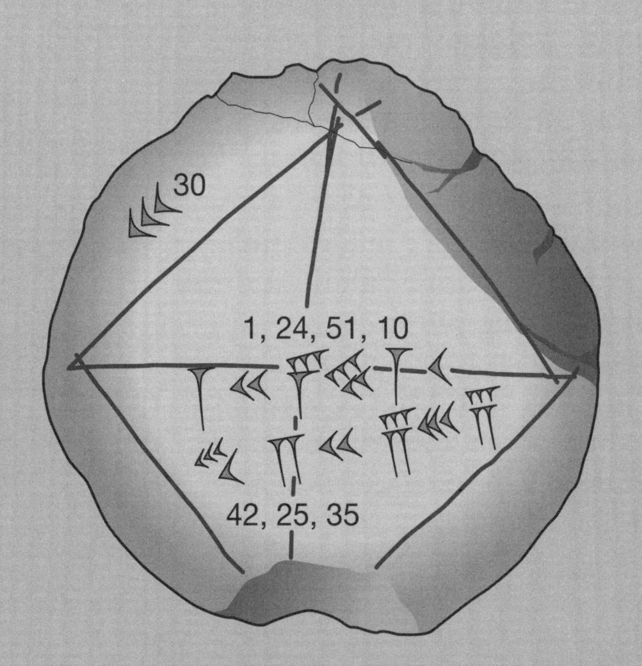

of the same type, suggesting that it may have been used for teaching purposes.

> *I found a stone, but did not weigh it; after I weighed out* 8 *times its weight, added* 3 *gin, and added one-third of one-thirteenth multiplied by* 21, *I weighed it:* 1 *ma-na. What was the original weight of the stone? The original weight was* $4^1/_3$ *gin.*

This problem is clearly not a practical one – if we want the weight of the stone, why don't we just weigh it? Unfortunately, we do not know how the scribe solved the problem – we just have the answer.

Our next example is more complicated, and is one of a dozen similar problems on the same tablet:

> *I have subtracted the side of my square from the area:* 14,30. *You write down* 1, *the coefficient. You break off half of* 1. 0;30 *and* 0;30 *you multiply. You add* 0;15 *to* 14,30. *Result* 14,30;15. *This is the square of* 29;30. *You add* 0;30, *which you multiplied, to* 29;30. *Result:* 30, *the side of the square.*

This is a quadratic equation: $x^2 - x = 870$, in modern algebraic notation. Here, x is the side of the square, x^2 is the area, and 14;30 is our decimal number 870. The steps of the above solution give, successively,

$1, {}^1/_2, ({}^1/_2)^2 = {}^1/_4, 870{}^1/_4, 29{}^1/_2, 30.$

The method in this example is called 'completing the square' and is essentially the one that we use today, 4000 years later.

THALES

Little is known about Thales (c.624–c.546BC). According to legend, he came from the Greek Ionian city of Miletus on the west coast of Asia Minor in modern-day Turkey. Many claims have been made for him: he visited Egypt and calculated the height of the pyramids, predicted a solar eclipse in 585BC, showed how rubbing feathers with a stone produces electricity, and originated the phrase 'know thyself'.

The Seven Sages of Greece: a woodcut from *The Nuremberg Chronicle* (1493); Thales is on the left.

Thales is widely considered the first important Greek mathematician. Bertrand Russell claimed that 'Western philosophy begins with Thales', and indeed Thales was considered one of the Seven Sages of Greece, a title awarded by tradition to seven outstanding Greek philosophers from the 6th century BC.

GREEK MATHEMATICAL SOURCES

Unlike Ancient Egypt, where there are few well-preserved papyri, and Mesopotamia where many thousands of clay tablets survive, we have very few Greek primary sources. As in Egypt, the Greeks wrote on papyrus which did not survive the centuries, and there were disasters, such as a library fire at Alexandria, in which many of the primary sources perished.

Thales of Miletus

So we have to rely mainly on commentaries and later versions. The best-known commentator on Greek mathematics was Proclus (5th century AD), who supposedly derived his material from earlier commentaries (now lost) by Eudemus of Rhodes (4th century BC). But Proclus lived 1000 years after Thales, so we should treat his commentaries with caution, while acknowledging that they are all we have.

GEOMETRY

The mathematical style developed by the early Greeks differed markedly from anything that went before. Of their many contributions to mathematics, and to geometry in particular, the ideas of deductive reasoning and mathematical proof are the most fundamental. Starting with some initial assumptions, known as *axioms* or *postulates,* they made simple deductions, then more complicated ones, and so on, eventually deriving a great hierarchy of results, each depending on previous ones.

THE THEOREMS OF THALES

A number of geometrical results have been ascribed to Thales by various commentators:

PROOF BY CONTRADICTION

In their geometry the Greeks used various methods of proof. For the following result, Thales gave a *proof by contradiction* (or *reductio ad absurdum*), where he assumed the desired result to be false and then deduced a consequence that contradicted this assumption — so the result is true.

ANY CIRCLE IS BISECTED BY ITS DIAMETER

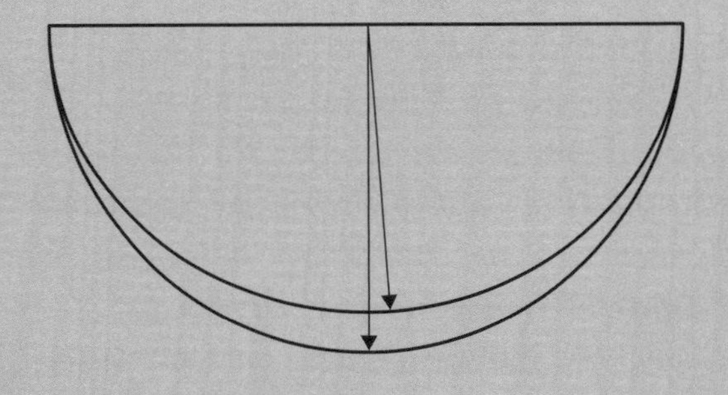

While writing about Euclid's *Elements*, Proclus remarked:

The famous Thales is said to have been the first to demonstrate that the circle is bisected by the diameter.

If you wish to demonstrate this mathematically, imagine the diameter drawn and one part of the circle fitted upon the other.

If it is not equal to the other, it will fall either inside or outside it, and in either case it will follow that a shorter line is equal to a longer. For all the lines from the centre to the circumference are equal, and hence the line that extends beyond will be equal to the line that falls short, which is impossible.

This contradiction proves the result.

The angle in a semicircle

If AB is a diameter of a circle, and if P is any other point on the circle, then the angle APB is a right angle.

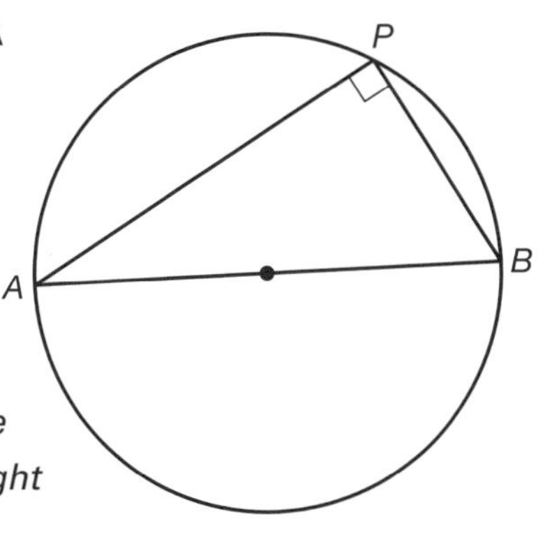

The intercept theorem

Let two lines intersect at a point P, and let two parallel lines cut these lines in the points A, B and C, D, as shown below. Then PA / AB = PC / CD.

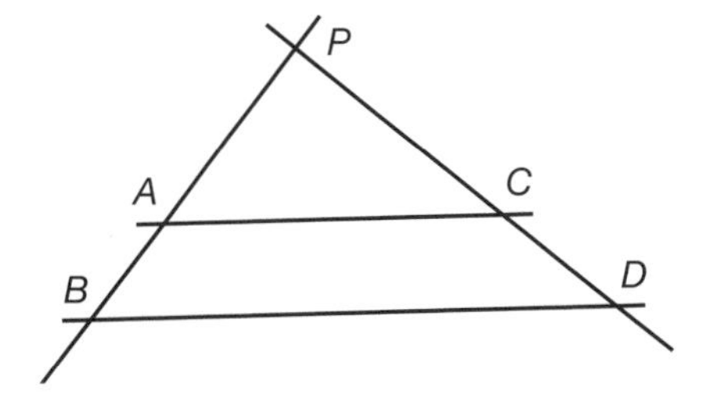

The base angles of an isosceles triangle

A triangle is *isosceles* if two of its sides are equal. The commentator Eudemus attributed to Thales the discovery that: *The base angles of an isosceles triangle are equal.*

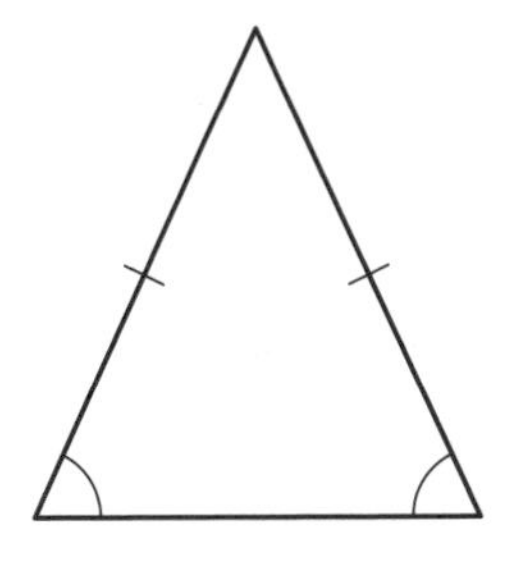

This last result was later known as the *pons asinorum* (Bridge of Asses). In medieval universities this was often as far as students ever reached: if you could cross the bridge of asses, you could then go on to all the treasures that lay beyond!

PYTHAGORAS

The semi-legendary figure of Pythagoras (c.570–490BC) was born on the island of Samos, in the Aegean Sea. In his youth he studied mathematics, astronomy, philosophy and music. Possibly around 520BC, he left Samos to go to the Greek seaport of Crotona (now in Southern Italy) and formed a philosophical school, now known as the *Pythagoreans.*

The inner members of the Pythagoreans (the *mathematikoi*) apparently obeyed a strict regime, having no personal possessions and eating only vegetables (except beans); the sect was open to both men and women.

The Pythagoreans studied mathematics, astronomy and philosophy. They believed that *everything is created from whole numbers,* and that anything worthy of study can be quantified. They are said to have subdivided the mathematical sciences into four parts: *arithmetic,*

Pythagoras, from Raphael's fresco *The School of Athens*

NUMBER PATTERNS

For the Pythagoreans, 'arithmetic' meant studying whole numbers, which they sometimes represented geometrically; for example, they considered *square numbers* as being formed by square patterns of dots or pebbles.

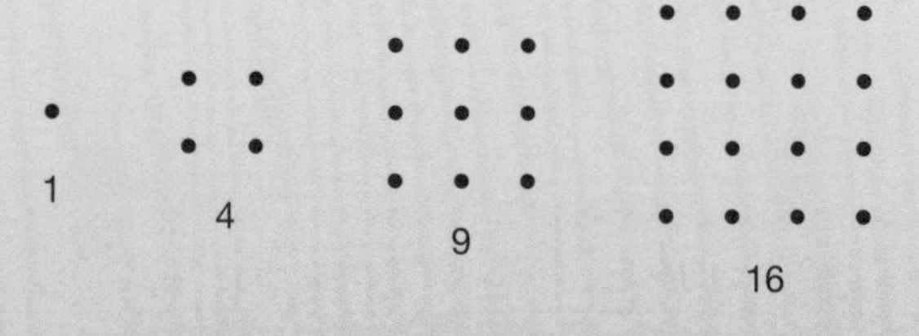

Using such pictures they could show that square numbers can be obtained by adding consecutive odd numbers, starting from 1 – for example, 16 = 1 + 3 + 5 + 7.

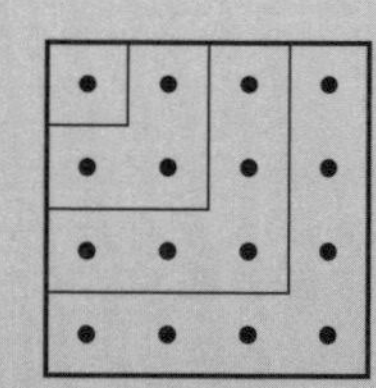

They also studied *triangular numbers,* formed by triangular patterns of dots. The first few triangular numbers are

1, 3, 6, 10, 15 and 21.

Notice that 3 = 1 + 2, 6 = 1 + 2 + 3, 10 = 1 + 2 + 3 + 4, etc.

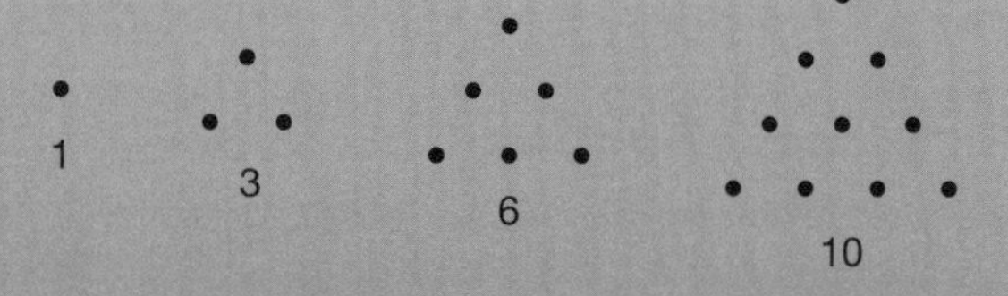

Using such pictures they could show that the sum of any two consecutive triangular numbers is a square number – for example, 10 + 15 = 25.

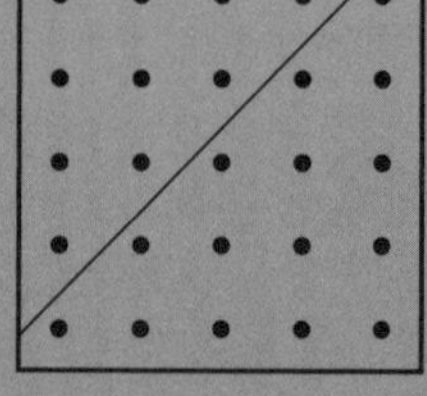

THE PYTHAGOREAN THEOREM

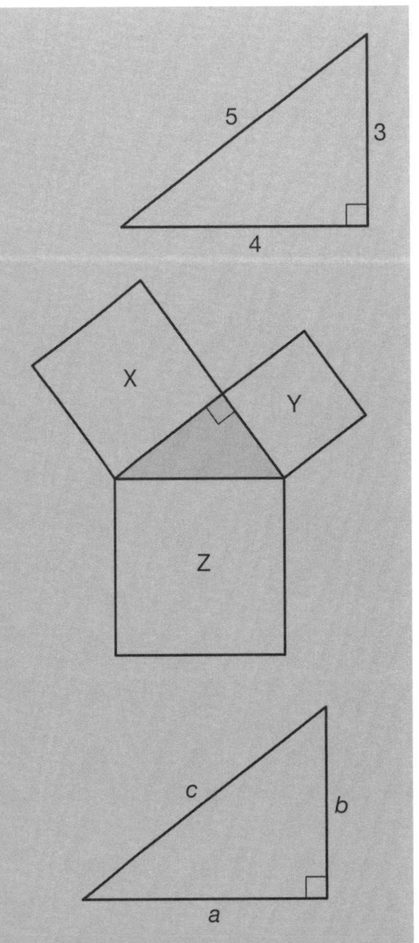

Important in geometry are *right-angled triangles,* where one of the angles is 90°; an example is the triangle with sides 3, 4, 5.

The most important result concerning them is known as the *Pythagorean theorem,* although no contemporary historical evidence links it to Pythagoras himself. Although it was known by the Mesopotamians 1000 years earlier, the Greeks were probably the first to prove it.

Geometrically, the Pythagorean theorem says that if we take a right-angled triangle and draw squares on each side of it, then *The area of the square on the longest side is equal to the sum of the areas of the squares on the other two sides* — that is, (area of Z) = (area of X) + (area of Y)

So, for a right-angled triangle with sides of lengths a, b and c (where c is the length of the longest side), we have $a^2 + b^2 = c^2$ — for example, for the triangle with sides 3, 4, 5,

$$3^2 + 4^2 = 9 + 16 = 25 = 5^2.$$

Other examples are the right-angled triangles with sides 5, 12, 13 and 8, 15, 17.

geometry, astronomy and *music* (later called the *quadrivium*). These subjects, in combination with the *trivium* (the liberal arts of grammar, rhetoric and logic), comprised the 'liberal arts' — the curriculum of academies and universities over the next 2000 years.

MATHEMATICS AND MUSIC

The Pythagoreans also experimented with music — in particular, linking certain musical intervals to simple ratios between small numbers.

It is likely that they discovered these ratios by plucking strings of different lengths and comparing the notes produced; for example, the harmonious interval of an *octave* results from halving the length of a string, giving a frequency ratio of 2 to 1, while another harmonious interval, a *perfect fifth,* results from stopping a string at two-thirds of its length, giving a ratio of 3 to 2.

A 1492 woodcut featuring some of Pythagoras's musical experiments.

PLATO AND ARISTOTLE

Plato and Aristotle, from Raphael's fresco *The School of Athens*

From about 500 to 300BC, Athens became the most important intellectual centre in Greece, numbering among its scholars Socrates, Plato (429–347BC) and Aristotle (384–322BC). Although neither is remembered primarily as a mathematician, both helped to set the stage for the 'golden age of Greek mathematics' in Alexandria.

PLATO'S ACADEMY

The next great era of Greek mathematics was centred on Athens, with the founding of Plato's Academy around 387BC in a suburb of Athens called 'Academy' (from which it derived its name). Here Plato wrote and directed studies, and the Academy soon became the focal point for mathematical and philosophical activities.

Plato believed that the study of these subjects provided the finest training for those who were to hold positions of responsibility in the state, and in his *Republic* he discussed at length the importance for the 'philosopher-ruler' of each of the four mathematical arts – arithmetic, geometry, astronomy and music. Significantly, over the entrance appeared the inscription:

Let no-one ignorant of geometry enter here.

THE PLATONIC SOLIDS

Plato's book *Timaeus* is also of mathematical interest and includes a discussion of the five regular solids:

tetrahedron, cube, octahedron,
dodecahedron and *icosahedron.*

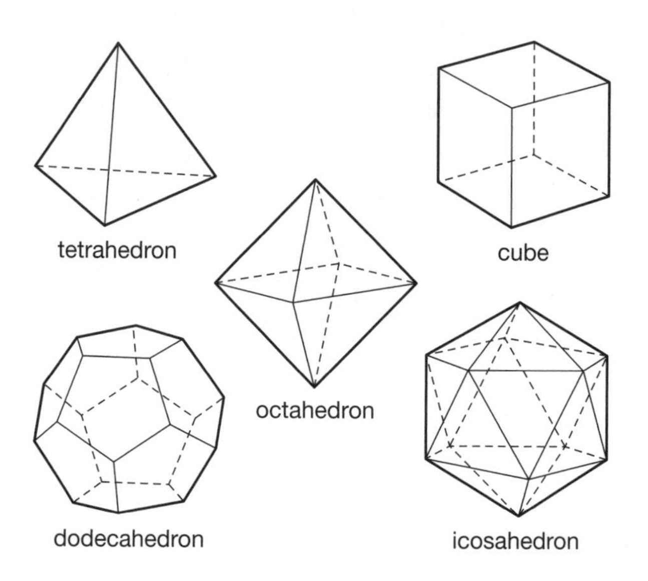

The five regular solids

In these solid figures (or *polyhedra,* meaning 'many-faced'), the faces are all regular polygons of the same type (triangles, squares or pentagons) and the arrangement of polygons at each corner is the same: for example, the cube has six square faces with three of them meeting at each corner, and the icosahedron has twenty triangular faces with five meeting at each corner.

In his *Timaeus* Plato linked the universe with the dodecahedron and assigned the other four polyhedra to the Greek elements of earth, air, fire and water. As a result, the regular polyhedra are often called *Platonic solids.*

SOCRATES AND THE SLAVE BOY

In his short dialogue *Meno,* Plato describes how Socrates drew in the sand a square of side 2 and area 4. He then asked a slave boy how to draw a square with double the area (8).

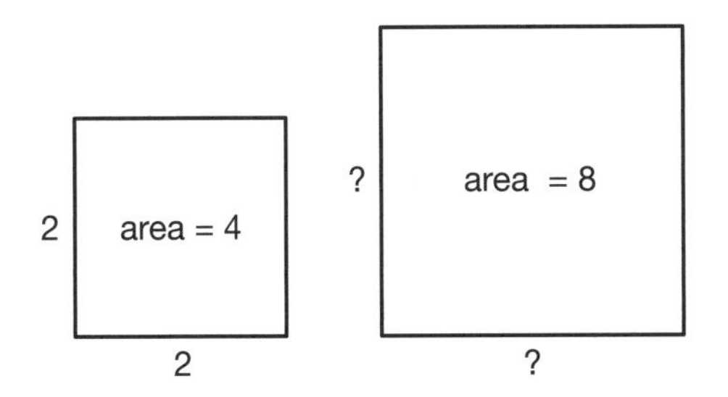

The boy first suggested doubling the side of the square to 4, but that gave four times the area (16). He then proposed a square with side 3, but that area was also too large (9). Eventually, after much discussion, he settled on the square based on the *diagonal* of the original square; this has area 8, as required.

The *Meno* dialogue is a wonderful example of teaching by asking questions, and was far removed from anything previously seen in Egypt or Mesopotamia.

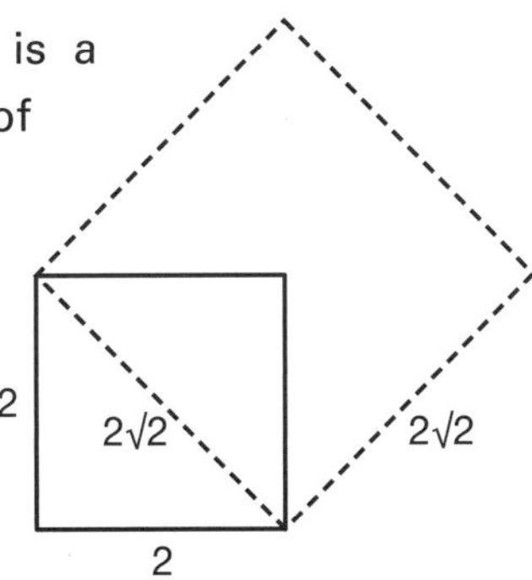

ARISTOTLE

Aristotle became a student at the Academy at the age of 17 and stayed there for some twenty years until Plato's death.

He was fascinated by logical questions and systematized the study of logic and deductive reasoning. In particular, he studied the nature of mathematical proof and considered deductions (known as *syllogisms*), such as

All men are mortal
Socrates is a man
Therefore Socrates is mortal

Aristotle also alluded to a proof that the ratio of the diagonal of a square to its side (that is, $\sqrt{2}$) cannot be written in fraction form p/q, where p and q are whole numbers.

$\sqrt{2}$ CANNOT BE WRITTEN AS A FRACTION p/q

The proof is by contradiction: we assume that $\sqrt{2}$ *can* be written as a fraction p/q, and show that this leads to a contradiction.

- *We can assume that this fraction is written in its lowest terms — that is, p and q have no common factor (other than 1).*
- *By squaring, we can rewrite the equation* $\sqrt{2} = p/q$ *as* $2 = p^2/q^2$, *and so* $p^2 = 2q^2$. *This means that* p^2 *must be an even number (because it is twice* q^2*), and so p must also be even (because if p were odd, then* p^2 *would also be odd).*
- *Since p is even, we can write* $p = 2k$, *for some whole number k. So* $p^2 = 2q^2 = 4k^2$, *which gives* $q^2 = 2k^2$. *It follows that* q^2 *is an even number, so q is also even.*
- *But this gives the required contradiction: p and q are both even, so both have a factor of 2. This contradicts the fact that p and q have no common factor.*

The contradiction arises from our original assumption that $\sqrt{2}$ can be written as a fraction p/q — so this assumption must be wrong: $\sqrt{2}$ cannot be written as a fraction.

EUCLID

Around 300BC, with the rise to power of Ptolemy I, mathematical activity moved to the Egyptian part of the Greek empire. In Alexandria Ptolemy founded a university that became the intellectual centre for Greek scholarship for over 800 years. He also started its famous library, which eventually held over half-a-million manuscripts before being destroyed by fire. Alexandria's Pharos lighthouse was one of the seven wonders of the ancient world.

THE *ELEMENTS*

The first important mathematician associated with Alexandria was Euclid (c.300BC), who is credited with writing on geometry, optics and astronomy. But he is mainly remembered for one work — the *Elements,* the most influential and widely read mathematical book of all time. It was in use for more than 2000 years and, apart from the Bible, may even be the most printed book ever.

Euclid's *Elements,* a model of deductive reasoning, was a compilation of known results organized in a logical order. Starting from initial axioms and postulates, it used rules of deduction to derive each new proposition in a systematic way. It was not the earliest such work, but was the most important.

It consists of thirteen sections, usually called 'Books' although they were written on rolls of papyrus. They are traditionally divided into three main parts — plane geometry, arithmetic and solid geometry.

PLANE GEOMETRY

The geometrical part (Books I to VI) opens with definitions of such basic terms as *point, line* and *circle,* followed by some axioms (or postulates) that permit us to carry out certain geometrical constructions with an unmarked ruler and a pair of compasses. These include:

- drawing a straight line from any given point to any other,
- drawing a circle with any given centre and radius.

Euclid then presented his first result, which gives a construction for an equilateral triangle (a triangle with all three sides equal):

Given a straight line AB, construct an equilateral triangle with AB as its base.

To do this, he used the second construction above to draw two circles, one with centre A and radius AB, and the other with centre B and radius AB. These circles meet at two points C and D, and the triangle ABC (or ABD) is then the required equilateral triangle.

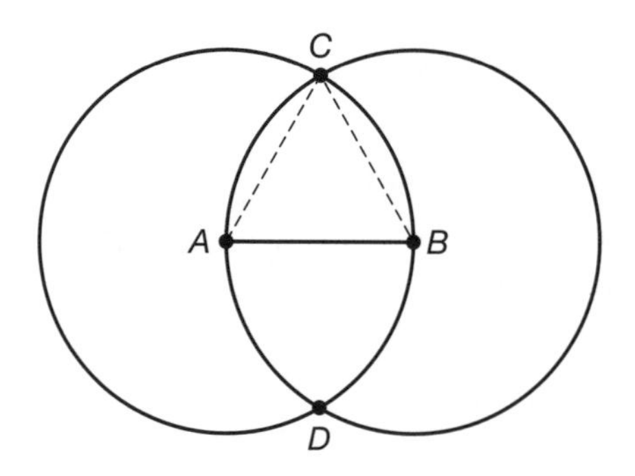

Euclid explained why this construction always gives an equilateral triangle. At each stage of his proof he made reference to an appropriate definition or postulate.

Book I continues with results about congruent triangles (those of the same size and shape) and parallel lines. Euclid also proved the 'angle-sum theorem',

The angles of any triangle add up to 180°,

and gave a proof of the Pythagorean theorem.

Book II includes various results on rectangles, such as the construction of a rectangle equal in

SOLID GEOMETRY

The final three books of Euclid's *Elements* deal with aspects of three-dimensional geometry. Of these, Book XIII is the most remarkable. Here, Euclid investigated the five regular solids (tetrahedron, cube, octahedron, dodecahedron and icosahedron) and showed how they can be constructed.

He concluded the *Elements* by proving that these are the only possible regular solids — there can be no others. This, the first ever 'classification theorem' in mathematics, forms a fitting climax to this great work.

area to a given triangle, while Book III introduces properties of circles, such as Thales' theorem on the angle in a semicircle, and a proof that

If a quadrilateral is drawn in a circle then the opposite angles add up to 180°.

$a + c = 180°$
$b + d = 180°$

Book IV contains circle constructions, including those of regular 5-sided, 6-sided and 15-sided polygons inside a given circle.

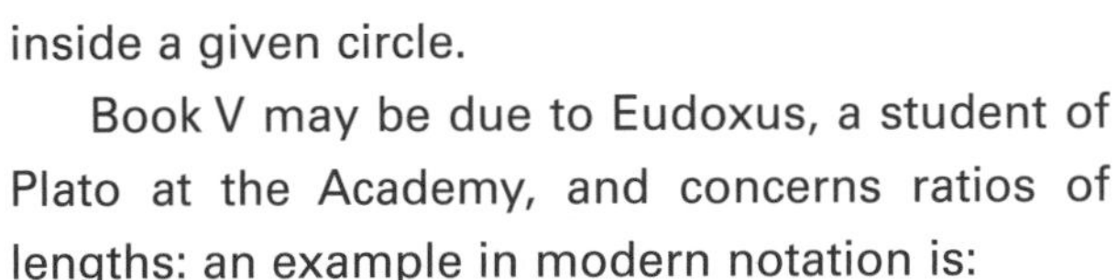

Book V may be due to Eudoxus, a student of Plato at the Academy, and concerns ratios of lengths: an example in modern notation is:

If $a / b = c / d$, then $a / c = b / d$.

These are then applied in Book VI to similar geometrical figures (those of the same shape but not necessarily the same size).

ARITHMETIC

In Books VII to IX, we enter the world of arithmetic, but the descriptions are still given in geometrical terms, using lengths of lines to represent numbers. There are discussions of odd and even numbers, and of what it means for one number to be a factor of another. Included here is the so-called *Euclidean algorithm,* a systematic method for finding the highest common factor of two numbers.

This section of the *Elements* also includes a discussion of prime numbers. A *prime number* is a number, greater than 1, whose only factors are itself and 1: the first few are

2, 3, 5, 7, 11, 13, 17, 19, 23 and 29.

They are central to arithmetic because they are the building blocks for numbers: every whole number can be obtained by multiplying prime numbers — for example,

$126 = 2 \times 3 \times 3 \times 7.$

Book IX contains Euclid's proof of the fact that the list of primes continues for ever:

There are infinitely many prime numbers.

It is one of the most famous proofs in the whole of mathematics.

Euclid presenting his *Elements* to King Ptolemy I Soter in Alexandria; the illustration is by Louis Figuier, 1866

ARCHIMEDES

Archimedes (c.287–212BC), a native of Syracuse on the island of Sicily and one of the greatest mathematicians of all time, worked over a wide range of areas. In geometry he calculated the surface areas and volumes of various solids, listed the 'semi-regular solids', investigated spirals and estimated the value of π. In applied mathematics he contributed to hydrostatics and discovered the law of the lever.

TWO STORIES

Archimedes is mainly known for two stories, recounted some two hundred years later and of doubtful authenticity.

The first was recalled by the Roman writer Vitruvius. Archimedes' friend King Hiero wished to find whether his crown was of pure gold or whether it was partially made of silver.

Archimedes, by Georg Andreas Böckler, 1661

Archimedes discovered the key to solving this problem while getting into a bath and observing that the more his body sank into it the more water flowed over the edge. Overjoyed at his discovery, he jumped out of his bath and ran home naked shouting 'Eureka!' (or more accurately 'Heureka!') — I have found it!

The other story, presented by Plutarch, concerns Archimedes' untimely death at the hand of a Roman soldier. In 212BC, during the siege of Syracuse, Archimedes was engrossed in a mathematical problem, unaware that the city had been captured, when the soldier came up and threatened to kill him. Archimedes begged him to wait until he had completed his calculations, whereupon the soldier flew into a rage and slew him on the spot.

TWO APPLICATIONS

Archimedes contributed to many areas of mathematics, and seems to be one of the few Greek mathematicians interested in its applications.

In hydrostatics, *Archimedes' principle* is that the weight of an object immersed in water is reduced by an amount equal to the weight of water displaced. Archimedes also devised ingenious mechanical weapons of war for the defence of Syracuse and is credited with inventing the Archimedean screw for raising water from a river.

Another result of his was the law of the lever — that if weights W_1 and W_2 are placed at the ends of a balance, then they are in equilibrium at distances a and b that are inversely proportional to the weights:

$$W_1 \times a = W_2 \times b.$$

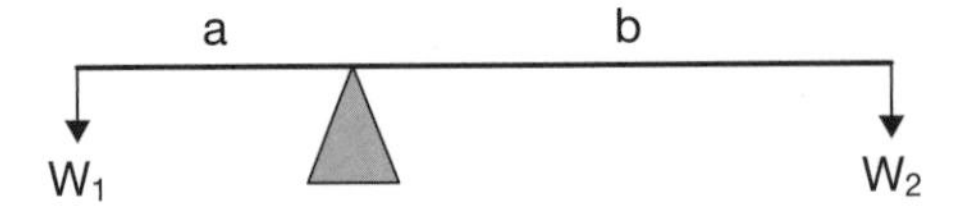

GEOMETRY

But Archimedes did not work only on applications. Among his best-known geometrical results are:

- determinations of the centres of gravity of triangles, parallelograms and hemispheres
- some work on spirals, including one now known as the *Archimedean spiral*
- impressive calculations of the volumes of spheres, cones and cylinders, such as his celebrated result (which he wanted engraved on his tomb) that *the volume of a cylinder is* $1\frac{1}{2}$ *times that of the sphere it surrounds.*

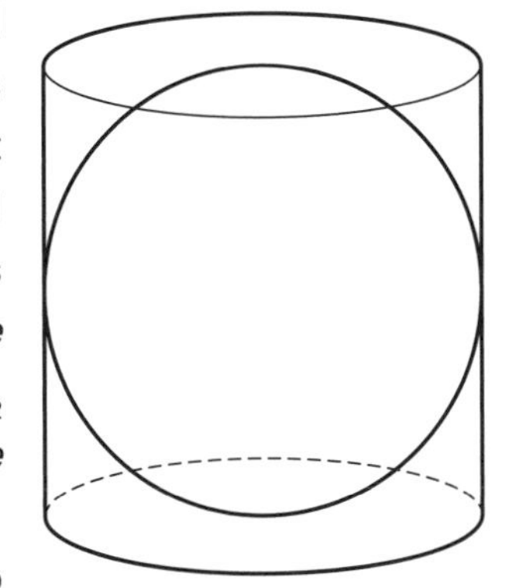

Archimedes also investigated the 'semi-regular' polyhedra, in which the faces are regular polygons but are not all the same; for example, a truncated icosahedron (or football) is made up of regular pentagons and hexagons. Archimedes found that there are just 13 such solids: they are now known as the *Archimedean polyhedra.*

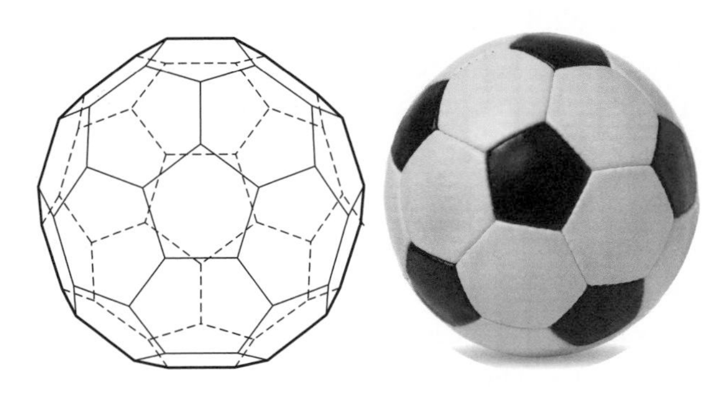

A truncated icosahedron and a football

COUNTING GRAINS OF SAND

In arithmetic, Archimedes wrote *The sand-reckoner* to refute the prevalent idea that the number of grains of sand in the universe is infinite. To this end, he first investigated the number $100{,}000{,}000^{100{,}000{,}000}$, which he called *P*, and then proceeded to construct the number $P^{100{,}000{,}000}$. As he carefully explained, this massive number is finite, yet exceeds the number of grains of sand in the universe. Since the Greek number system had names for numbers up to a myriad (10,000) and no further, this was a remarkable achievement.

CIRCLE MEASUREMENT

One of Archimedes' best-known results concerns the ratio of a circle's circumference to its diameter (that is, π). He began by drawing hexagons inside and outside a circle and compared their perimeters with the circumference of the circle: this tells us that π lies between 3 and 3.464.

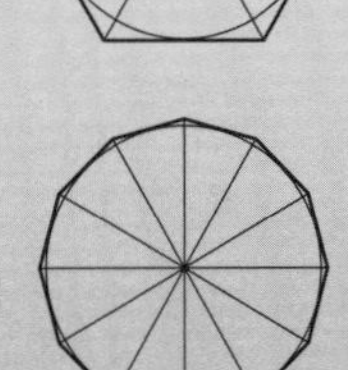

He then replaced the hexagon by a 12-sided polygon and recalculated the lengths.

Continuing in this way, and performing the calculations for polygons with 24, 48 and 96 sides, without actually drawing them, he concluded that (in our notation) $3\frac{10}{71} < \pi < 3\frac{1}{7}$, — this gives a value for π of about 3.14, which is correct to two decimal places.

As we shall see, this method was refined over the next 2000 years to yield the value of π to many decimal places.

APOLLONIUS

Back in Alexandria, Apollonius of Perga (c.262–c.190BC), known since antiquity as 'the Great Geometer', was writing his celebrated treatise on conics: these curves are of three different types – an ellipse (with a circle as a special case), a parabola and a hyperbola. Apollonius's *Conics* was a veritable *tour de force,* but it is not an easy work to read.

THREE TYPES OF CONIC

The conic sections are generally considered to have been discovered by Menaechmus, a pupil of Eudoxus. By slicing a cone in various ways, he could obtain the following curves:

- slicing it horizontally gives a *circle*
- slicing it at a slant gives an *ellipse*
- slicing it parallel to the side of the cone gives a *parabola*
- slicing it vertically gives a *hyperbola*

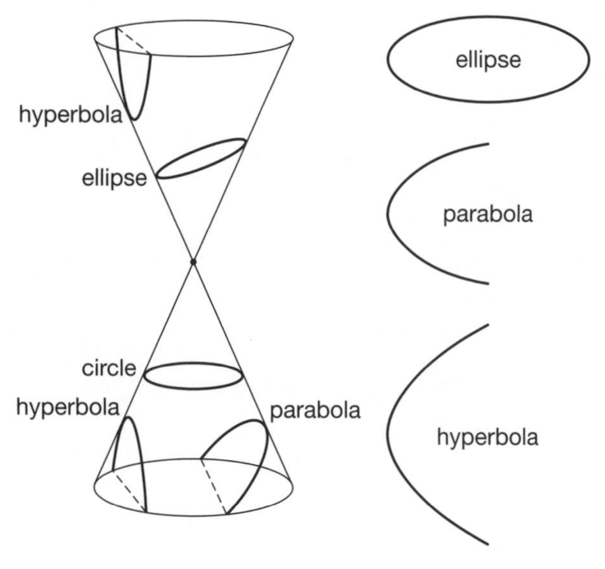

Slicing a cone

These curves can also be obtained by taking a fixed point (the *focus*) and a fixed line (the *directrix*) and letting a point P move so that its distances from them are in a fixed ratio r. For $r = 1$ we have a parabola; for $r < 1$ we have an ellipse; for $r > 1$ we have a hyperbola.

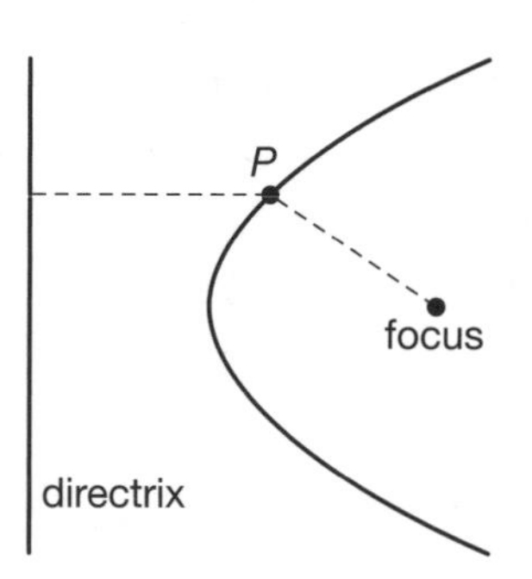

The focus–directrix definition of a parabola

Another way to draw an ellipse (used by gardeners to make elliptical flower beds) is to tie a rope to two pegs and trace out the curve.

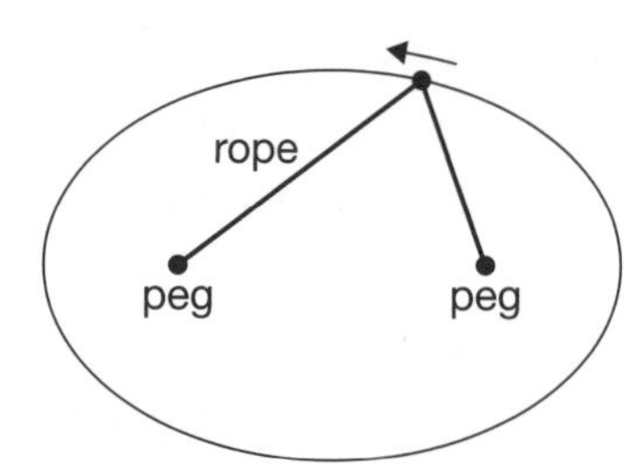

APOLLONIUS'S *CONICS*

While still a young man, Apollonius travelled to Alexandria to study with the followers of Euclid. He remained there, teaching and writing a number of geometrical works, the most influential of which was his monumental treatise on *Conics*. This appeared in eight parts: the early ones contain basic material, much of it previously known, while later ones present some stunningly original results.

Most of what we know about Apollonius's life appears in the letters that preface these parts: here we learn that he made visits to Pergamum and Ephesus to discuss his work with fellow geometers.

TWO LATER EDITIONS OF APOLLONIUS'S CONICS

After the invention of printing in the 15th century, many Greek writings appeared in book form. Here we see a 16th-century edition of the *Conics*, and the frontispiece of a 1710 edition by Edmond Halley (of 'Halley's comet' fame). The latter depicts the Greek philosopher Aristippus, shipwrecked with his fearful colleagues on the island of Rhodes; on noticing some geometrical figures drawn in the sand, Aristippus exclaimed 'Let us be of good cheer, for I see the traces of man'.

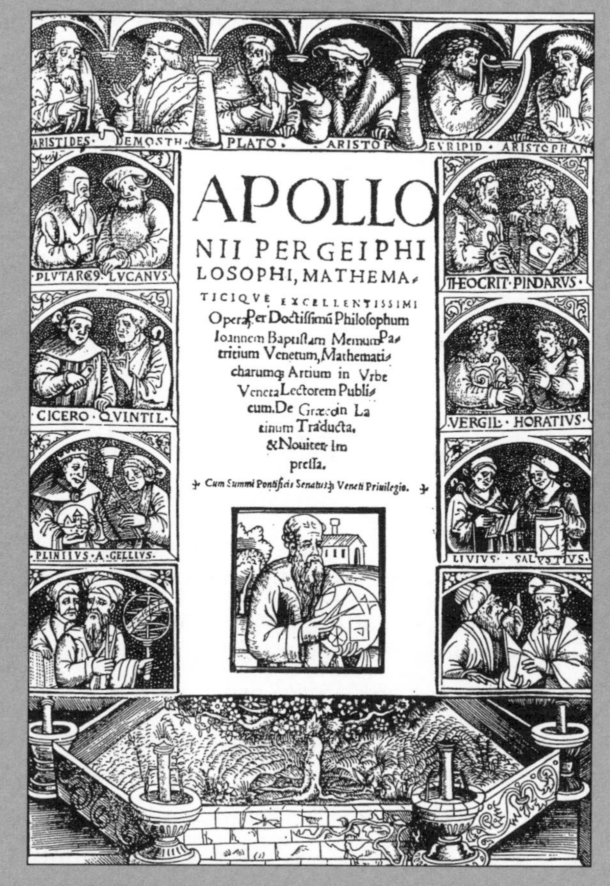
ARISTIDES · DEMOSTH · PLATO · ARISTOT · EVRIPID · ARISTOPHAN

APOLLO NII PERGEI PHI LOSOPHI, MATHEMA TICIQVE EXCELLENTISSIMI Opera Per Doctissimũ Philosophum Joannem Baptistam Memum Patritium Venetum, Mathematicharumq; Artium in Vrbe Veneta Lectorem Publicum. De Græco in Latinum Traducta, & Nouiter Impressa.

Cum Summi Pontificis Senatusq; Veneti Priuilegio.

PLVTARC9 · LVCANVS · THEOCRIT · PINDARVS · CICERO · QVINTIL · VERGIL · HORATIVS · PLINIVS · A · GELLIVS · LIVIVS · SALVSTIVS

A 16th-century edition

Halley's 1710 edition

THE CIRCLE OF APOLLONIUS

One of Apollonius's most famous results is known as the *circle of Apollonius:*

> *Suppose that a point P moves in the plane so that its distance from a point A is in a fixed ratio (≠ 1) to its distance from a point B. Then the point traces out a circle.*

The diagram opposite shows the circle traced out when the point P is always twice as far from A as it is from B.

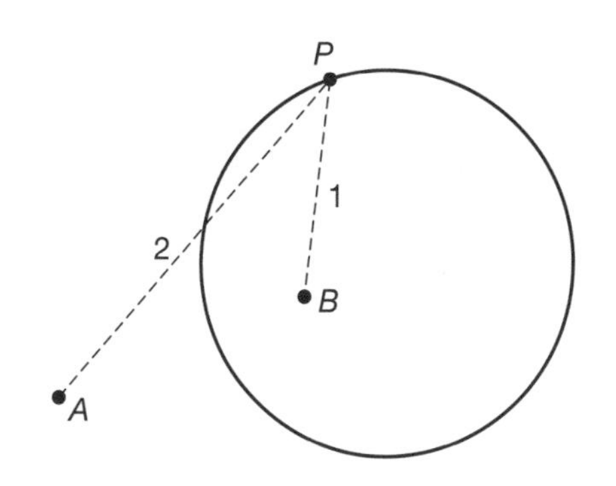

The circle of Apollonius

HIPPARCHUS AND PTOLEMY

The first trigonometrical approach to astronomy was made by Hipparchus (190–120BC), sometimes called 'the Father of Trigonometry'. Possibly the greatest astronomical observer of antiquity, he discovered the precession of the equinoxes, produced the first known star catalogue, and constructed a 'table of chords' yielding the sines of angles. Claudius Ptolemy of Alexandria (AD c.100–c.170) built on the work of Hipparchus and others to produce his great work on astronomy, commonly known as the *Almagest*.

HIPPARCHUS

Although born in Nicaea in Bithynia, Hipparchus spent most of his life in Rhodes, where he built on the observations of earlier Greek astronomers and Babylonian records to produce a fine star catalogue and set of planetary observations.

His use of these records and his own observations also led to what many claim as his finest achievement: his discovery of the precession of the equinoxes from a consideration of the slow motion of the points of the equinox and the solstice in the fixed stars. He also classified stars by their brightness, using a scale that ran from 1 (the brightest) to 6 (the dimmest).

Hipparchus gazing at the stars, in an engraving from J. N. Larned's *History of the World*, Vol. 1 (1897)

Hipparchus incorporated data from his astronomical observations into his geometrical models used to explain astronomical motions. He may also have developed an instrument of astrolabe type to derive the time at night from stellar observations.

SOME TRIGONOMETRY

Although little of Hipparchus's work survives, Claudius Ptolemy considered him his most important predecessor. Indeed, the subject of *trigonometry* (meaning angle-measuring), introduced by Hipparchus around 150BC, was developed by Claudius Ptolemy.

Fundamental to their work in astronomy was the calculation of the lengths of chords of circles — a chord is a line joining two points on the circle; it corresponds to working out the trigonometrical ratio called the sine for various angles.

The trigonometrical ratios arise from the study of right-angled triangles. If θ is the angle shown, we define the *sine, cosine* and *tangent* of θ (written $\sin\theta$, $\cos\theta$ and $\tan\theta$) by the following ratios of lengths:

$\sin\theta$ = opposite side / hypotenuse = a/c
$\cos\theta$ = adjacent side / hypotenuse = b/c
$\tan\theta$ = opposite side / adjacent side = a/b

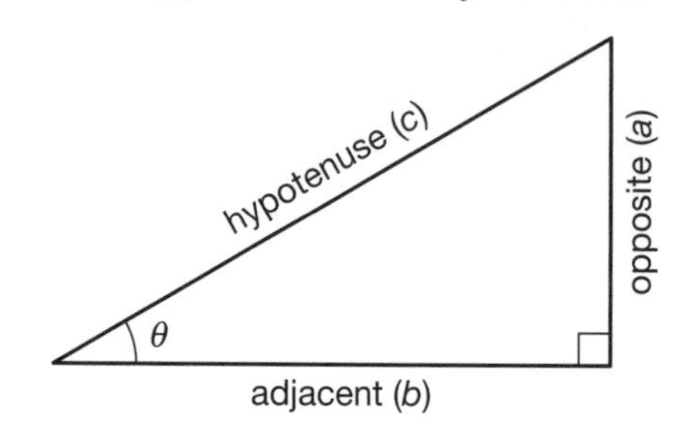

Ptolemy with his cross-staff for measuring the heavens in this drawing by André Thevet, 1584

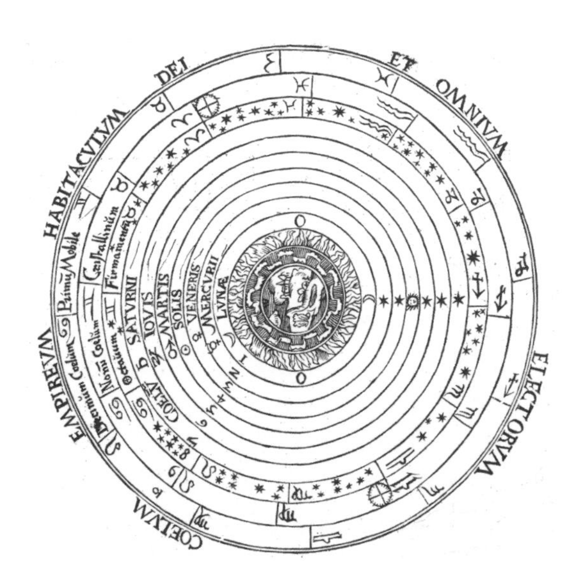

The Ptolemaic system, as portrayed in a Latin edition of the *Almagest*

To describe the motion of the sun and planets, Ptolemy introduced *epicycles*, small circles centred on the main circular orbit on which the sun or a planet is seen to move. Appropriately adjusting distances, the centre of rotation and the rates of rotations enabled him to make his accurate predictions.

PTOLEMY'S *ALMAGEST*

Ptolemy's definitive 13-volume work on astronomy, the *Syntaxis*, is usually known by its later Arabic name *Almagest* (The Greatest). Dominating astronomy for nearly 1500 years, it contains a mathematical description of the motion of the sun, moon and planets, and has a table of chords equivalent to listing the sines of angles from 0° to 90° in steps of $\frac{1}{4}$°.

The *Almagest* is our most important source of information on Hipparchus, and we also know little of Ptolemy beyond what appears there.

Ptolemy's astronomical observations relate to the period AD127–141 and were based at Alexandria; this is why he is often known as Claudius Ptolemy of Alexandria.

The *Almagest* developed a geometrical theory that could predict the motion of the planets to extraordinary accuracy. Ptolemy's geocentric cosmology regarded the earth as fixed and unmoving, with the sun and planets rotating around it.

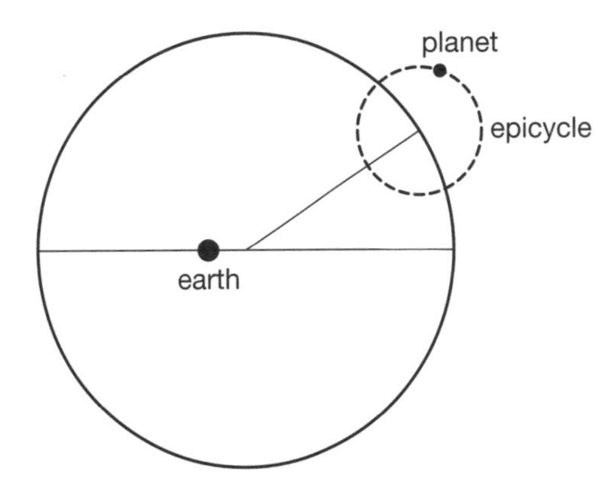

An epicycle in the Ptolemaic system

PTOLEMY'S *GEOGRAPHIA*

Ptolemy also published a standard and influential work on map-making called *Geographia*, in which he discussed various types of map projection and listed the latitude and longitude of 8000 places in the known world. His findings were used by navigators for over 1500 years.

DIOPHANTUS

Diophantus of Alexandria, known as 'the Father of Algebra', probably lived in the 3th century AD. We know little about his life. His main contributions to mathematics were the 13 books that comprise his *Arithmetica,* not all of which survive. Unlike the geometrical writings of most Greek mathematicians this was a collection of algebraic problems that were posed and solved. Diophantus was also the first mathematician to devise and employ algebraic symbols. As we shall see, his *Arithmetica* was to have a great influence over the coming centuries.

DIOPHANTINE PROBLEMS

Suppose that you wished to solve the equation $3x + y = 10$.

If x and y are allowed to take any numerical values, then there are infinitely many possibilities for them — for example:

- if $x = 1$, then $y = 7$
- if $x = -2\frac{1}{3}$, then $y = 17$
- if $x = \pi$, then $y = 10 - 3\pi$

But if x and y are both required to be positive whole numbers, then only three solutions are possible:

$$x = 1,\ y = 7; \quad x = 2,\ y = 4; \quad \text{and} \quad x = 3,\ y = 1.$$

An algebraic equation where restrictions are imposed on the solutions (usually, that they are whole numbers) is called a *Diophantine equation,* although Diophantus himself was often content to accept solutions that are fractions.

HOW OLD WAS DIOPHANTUS?

The following problem appears in a 5th-century collection known as *The Greek Anthology.*

This tomb holds Diophantus.
Ah, how great a marvel!
The tomb tells scientifically
the measure of his life.
God granted him to be a boy
for the sixth part of his life,
and adding a twelfth part to this,
He clothed his cheeks with down;
He lit him the light of wedlock
after a seventh part,
and five years after his marriage
He granted him a son.
Alas! Late-born wretched child;
after attaining the measure of half
his father's life, chill Fate took him.
After consoling his grief by this science
of numbers for four years, he ended his life.

What can we deduce from this?

The problem tells us that Diophantus spent $\frac{1}{6}$ of his life in childhood, $\frac{1}{12}$ of it as a youth, and $\frac{1}{7}$ more as a bachelor. Five years after his marriage there was a son who died four years before his father, when $\frac{1}{2}$ his father's final age. With modern algebraic notation, if x is Diophantus's age when he died, we have the equation

$$(\tfrac{1}{6}x + \tfrac{1}{12}x + \tfrac{1}{7}x) + 5 + \tfrac{1}{2}x + 4 = x.$$

Solving this equation gives $x = 84$, so Diophantus's life-span was 84 years.

DIOPHANTUS'S NOTATION

The word *arithmetic* is derived from the Greek word *arithmos,* meaning number. For their decimal counting system the Greeks used the 24 letters of their alphabet and three archaic symbols, as follows:

1	2	3	4	5	6	7	8	9
α	β	γ	δ	ε	ς	ζ	η	θ

10	20	30	40	50	60	70	80	90
ι	κ	λ	μ	ν	ξ	ο	π	ϙ

100	200	300	400	500	600	700	800	900
ρ	σ	τ	υ	φ	χ	ψ	ω	ϡ

For example, 648 was written as χμη. Sometimes these numbers would have bars over them to distinguish them from letters.

For his algebraic equations, Diophantus introduced further symbols, such as K^u for a cube, Δ^u for a square, ς for a first power and ° for addition, and would write a quadratic expression such as $2x^2 + 3x + 4$ as something like $\Delta^u \beta°ς\gamma°\delta$

SOME PROBLEMS OF DIOPHANTUS

Here we present some problems from the *Arithmetica.* Diophantus did not present general methods for solving his problems, but often chose a particular example and found the result in that case alone. The following solutions are adapted from the *Arithmetica*: note that he was quite happy to calculate with negative numbers.

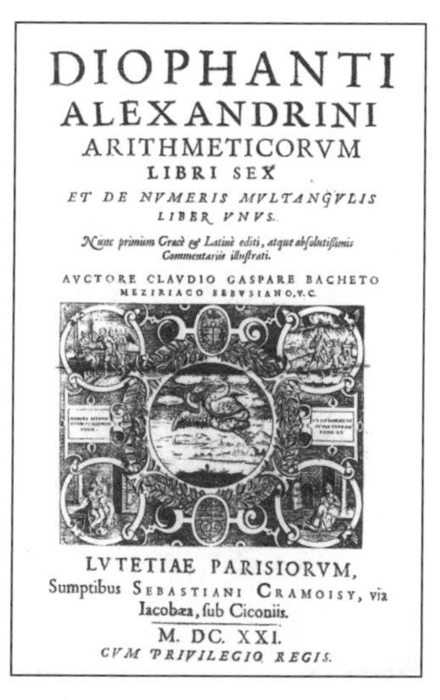
DIOPHANTI ALEXANDRINI ARITHMETICORVM LIBRI SEX ET DE NVMERIS MVLTANGVLIS LIBER VNVS.
Nunc primum Græcè & Latinè editi, atque absolutissimis Commentariis illustrati.
AVCTORE CLAVDIO GASPARE BACHETO MEZIRIACO SEBVSIANO, V.C.
LVTETIAE PARISIORVM, Sumptibus SEBASTIANI CRAMOISY, via Iacobæa, sub Ciconiis.
M. DC. XXI.
CVM PRIVILEGIO REGIS.

A 1621 French edition of Diophantus's *Arithmetica*

To find two numbers such that their sum and product are given numbers.
Given sum 20, *given product* 96.
Let $2x$ *be the difference of the numbers. Therefore the numbers are* $10 + x$, $10 - x$.
[Note that 10 is half of the sum, 20.]
Hence, $100 - x^2 = 96$.
So $x = 2$, *and the numbers are* 12 *and* 8.

To divide a given square into two squares. Let it be required to divide 16 *into two squares. And let the first square* = x^2;
then the other will be $16 - x^2$. *It is required therefore to make* $16 - x^2$ = *a square.*

I take a square of the form $(mx - 4)^2$, *m being any integer and* 4 *the root of* 16; *for example, let the side be* $2x - 4$ *and the square itself be* $4x^2 + 16 - 16x$. *Then* $4x^2 + 16 - 16x = 16 - x^2$. *Add to both sides the negative terms and take like from like. Then* $5x^2 = 16x$, *and* $x = {}^{16}/_{5}$.
One number is therefore ${}^{256}/_{25}$, *the other* ${}^{144}/_{25}$, *and their sum is* 16 *and each is a square.*

Many of his problems were even more complicated, and a few were described in geometrical terms:

To find a right-angled triangle whose perimeter is a square, while its perimeter added to its area gives a cube.

His answer was the triangle with sides

${}^{1024}/_{217}$, ${}^{215055}/_{47089}$ and ${}^{309233}/_{47089}$

PAPPUS AND HYPATIA

The Neoplatonists were religious and mystical philosophers from the 3rd century AD and later. Founded by the philosopher Plotinus, the school based its teachings on the works of Plato and his followers. Two important members were the Alexandrian geometers Pappus (c.290–c.350) and Hypatia (c.360–415).

PAPPUS

Pappus of Alexandria was one of the last Greek mathematicians of antiquity. We know little about his life, except that in his commentary on Ptolemy's *Almagest* he remarked on seeing a solar eclipse in Alexandria; from this we can deduce that he flourished around the year 320.

Pappus's most important work was his eight-volume *Mathematical Collection,* which surveyed a wide range of mathematical topics in arithmetic, plane and solid geometry, astronomy and dynamics. Here we look at two of his geometrical contributions.

ON THE SAGACITY OF BEES

There are only three ways of tiling a large floor with regular polygons of the same type. These are the patterns of squares, equilateral triangles and regular hexagons.

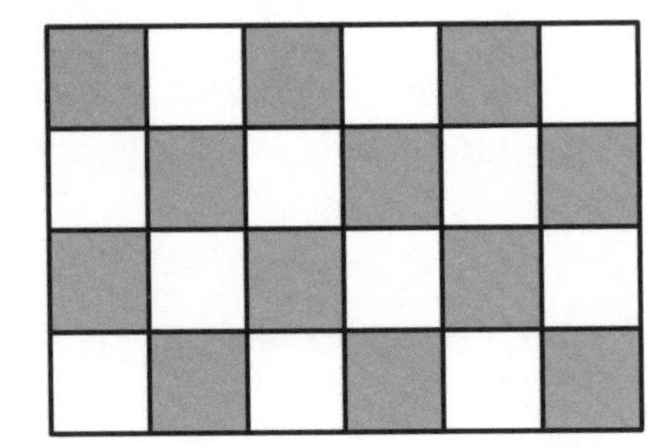

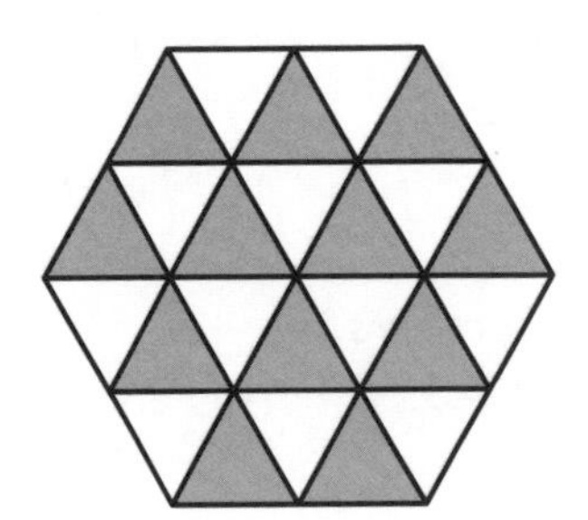

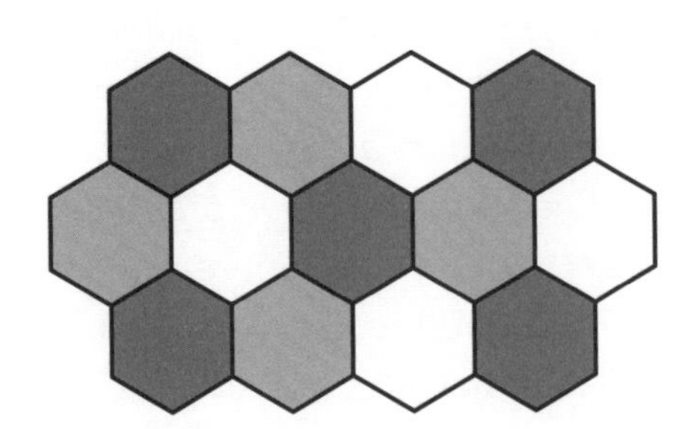

In Book V of his *Collection,* Pappus credited bees with a certain geometrical forethought in planning their honeycombs. After showing that there can only be the three regular arrangements above, he noted that the bees in their wisdom choose the pattern that has the most angles, the hexagon, perceiving that it holds more honey than the other two.

HYPATIA OF ALEXANDRIA

The first important woman mathematician that we know of was Hypatia, daughter and pupil of the geometer Theon of Alexandria. A noted geometer herself, she became Head of the Neoplatonic school in Alexandria around the year 400, and was apparently such a renowned expositor and lecturer that people came many miles to hear her.

Hypatia is credited with impressive commentaries on many classic texts, such as Apollonius's *Conics* and Diophantus's *Arithmetic,* and with an edition of Ptolemy's *Almagest.* She also demonstrated the construction of astronomical and navigational instruments, such as the astrolabe.

Tragically, her life was savagely cut short in the year 415 when she suffered a gruesome death at the hands of a fanatical mob of religious zealots opposed to Neoplatonism. Her murder was a death blow to mathematics in Alexandria.

Hypatia has been immortalized in several

PAPPUS'S THEOREM

Another celebrated result of Pappus is one of the great theorems of mathematics. It concerns the intersections of lines in a plane and is called the 'hexagon' theorem.

THE 'HEXAGON' THEOREM

Draw two straight lines on a sheet of paper, and choose any three points *A*, *B*, *C* on the first line, and any three points *P*, *Q*, *R* on the second line.

Now draw the lines

AQ* and *BP* — these meet at a point *X
AR* and *CP* — these meet at a point *Y
BR* and *CQ* — these meet at a point *Z

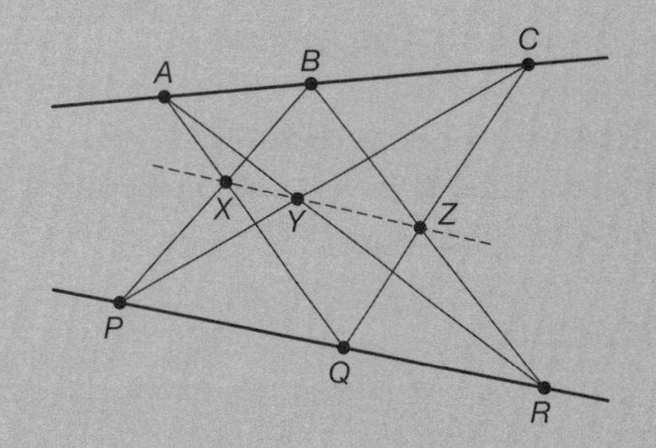

Pappus's theorem states that, however we choose our initial six points,

The resulting points X, Y and Z always lie on a straight line.

paintings and in literature. In 1853 the Victorian writer Charles Kingsley, best known for his fairy tale *The Water-Babies* and his historical novels *Westward Ho!* and *Hereward the Wake,* wrote *Hypatia,* a fictionalized biography set in 5th-century Alexandria.

Hypatia was murdered by a Christian mob in AD415, as portrayed by Louis Figuier in *Les Vies des Savants* (1875)

NICOMACHUS AND BOETHIUS

Nicomachus of Gerasa (c.AD60–120) wrote influential texts on arithmetic and music based on the work of Pythagoras. Later, Boethius (c.480–524) produced similar books on arithmetic and geometry. These works were all in constant use for hundreds of years.

NICOMACHUS

As a follower of Pythagoras, Nichomachus included in his *Introduction to Arithmetic* much that is familiar from what we know of Pythagoras's work. Here we find lengthy accounts of odd, even, square and triangular numbers. Nicomachus developed these ideas further, exploring pentagonal, hexagonal, heptagonal and tetrahedral numbers. Also included is the first known multiplication table featuring Greek numerals.

Clockwise from top left: Boethius, Pythagoras, Nicomachus and Plato, from a medieval manuscript

An illustration from Gregorius Reisch's encyclopedia *Margarita Philosophica* (1503), featuring the quadrivium subject of *Arithmetic*; the drawing contrasts the new arithmetic (represented by Boethius and Hindu–Arabic numerals) with the old (Pythagoras's counting board)

Nicomachus also discussed prime and perfect numbers. A number is *perfect* if it equals the sum of its factors (apart from itself); for example, $28 = 1 + 2 + 4 + 7 + 14$ is perfect. In his *Elements,* Euclid gave a formula that yields such numbers, and Nicomachus listed four of them, 6, 28, 496 and 8128.

BOETHIUS

In comparison with the Greeks, the Romans contributed little to the development of mathematics, using it mainly for practical concerns, such as architecture, surveying and adminstration.

Boethius came from a Roman family and, although orphaned at a young age, was well

Another quadrivium illustration from *Margarita Philosophica,* featuring *Astronomy*: Ptolemy is pictured wearing a crown as he was regularly confused during the Renaissance with the Egyptian King Ptolemy

educated and spent his life writing and translating. He was an enthusiast for the Greek quadrivium, the mathematical arts of arithmetic, geometry, astronomy and music.

Although his knowledge of mathematics was somewhat meagre, he wrote an *Arithmetic* in Latin, deriving much from Nicomachus's text, and a *Geometry,* based on results from the first four Books of Euclid's *Elements.*

The writings of both authors leave much to be desired. Nicomachus's *Arithmetic* contains many errors and omits all proofs, and Boethius's works were equally pedestrian. In spite of this, they were the standard texts in these subjects for many hundreds of years, at a time when little else was happening in mathematics.

THE CHINESE

China's mathematical history dates back 3000 years or more. Around 220BC the ancient Chinese built the Great Wall, a major triumph of engineering skill and mathematical calculation. The Chinese may have been the first to develop a decimal place-value system, similar to the one we use today; they also constructed sundials and were early users of the abacus.

MAGIC SQUARES

An ancient Chinese legend concerns Emperor Yu of Xia, who was standing on the banks of the river Lo (a tributary of the Yellow River) when a sacred turtle emerged from the river with the numbers 1 to 9 on its back. These numbers appeared in the form of a 3 × 3 *magic square* (the *lo-shu*), an arrangement of numbers in which the numbers in each row, column and diagonal have the same sum:

$$4 + 9 + 2 = 9 + 5 + 1 = 4 + 5 + 6 = 15, \text{ etc.}$$

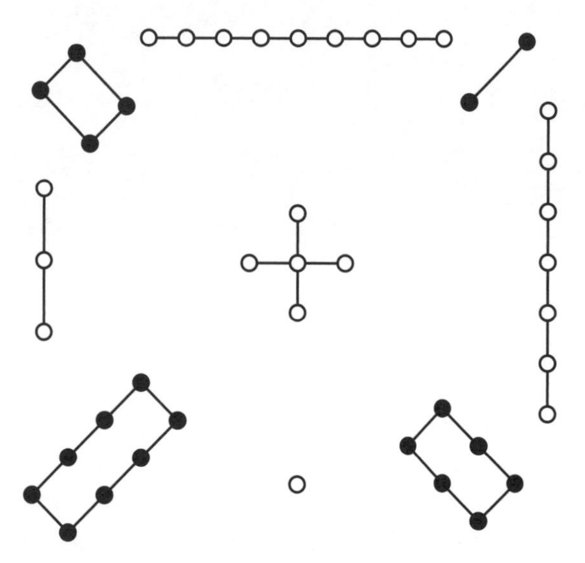

This particular arrangement of numbers acquired great religious and mystic significance over the centuries and appeared in many forms. Although Emperor Yu lived around 2000BC, no account of this story appeared until much later – possibly as late as the Han dynasty, which began in 206BC.

COUNTING

For their calculations the Chinese used a form of counting board, a box with separate compartments for units, tens, hundreds, etc., into which small bamboo rods were placed. Each symbol from 1 to 9 had two forms, horizontal and vertical, enabling the calculator to distinguish easily between the numbers in adjacent compartments. Here are the numbers 1713 and 6036.

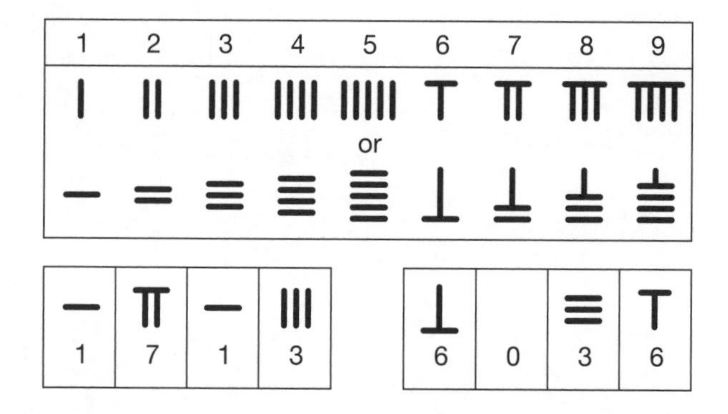

GOU-GU (PYTHAGOREAN THEOREM)

The Chinese used the idea of dissection (cutting up and reassembling) to obtain results in geometry. A celebrated example is *gou-gu*, their name for the Pythagorean theorem, which

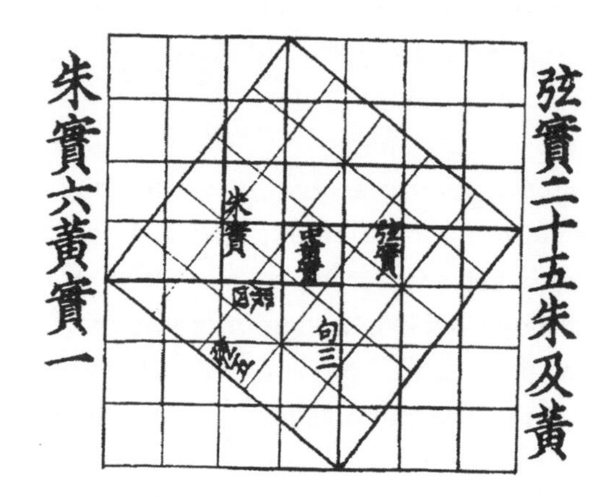

CIRCLE MEASUREMENT

Several Chinese mathematicians devoted their energies to estimating π.

Around the year AD100 Zhang Heng, inventor of the seismograph for measuring the intensity of earthquakes, proposed the value $\sqrt{10}$ (about 3.16 in our decimal notation).

We saw how Archimedes calculated with regular polygons having 6, 12, 24, 48 and 96 sides to obtain the estimates $3^{10}/_{71} < \pi < 3^{1}/_{7}$ (about 3.14 in our decimal notation). In his *Haidao suanjing* (Sea Island Mathematical Classic) of 263, Liu Hui continued to double the number of sides until he reached polygons with 3072 sides, obtaining the value $\pi = 3.14159$.

The Chinese fascination with π reached its climax in the 5th century, when Zu Chongzhi and his son calculated the areas of polygons with 24,576 sides and deduced that

$3.1415926 < \pi < 3.1415927$.

They also found the estimate of $3^{16}/_{113}$ (= $^{355}/_{113}$), which gives π to six decimal places: this approximation was not rediscovered in Europe until the 16th century.

appears in the *Zhou bi suan jing* (Mathematical Classic of the Zhou Gnomon) from before 100BC. Our explanation below uses modern algebraic notation.

The diagram shows a tilted square (of side c, say), surrounded by four right-angled triangles (with sides a, b and c), making a large square (of side $a + b$). We now cut this large square of side $a + b$ into five pieces: the square of side c and the four triangles, each with area $^{1}/_{2}ab$. Then the area of the large square is $c^2 + (4 \times {}^{1}/_{2}ab) = c^2 + 2ab$ and is also

$(a + b)^2 = a^2 + b^2 + 2ab$ – so $a^2 + b^2 = c^2$.

BAMBOO PROBLEM

A classic Chinese problem is the problem of the broken bamboo; a chi is a unit of length:

> *A bamboo 10 chi high is broken, and the upper end reaches the ground 3 chi from the stem. Find the height of the break.*

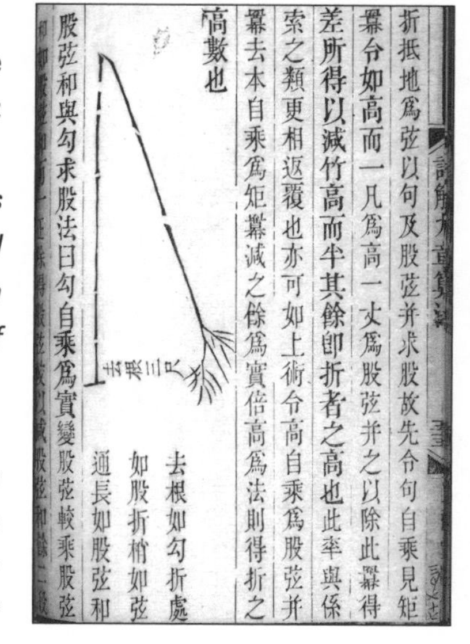

折抵地爲弦以句及股弦并求股故先令句自乘見矩
冪令如高而一凡爲高一丈爲股弦并之以除此冪得
差所得以減竹高而半其餘即折者之高也此率與係
索之類更相返覆也亦可如上術令高自乘爲股弦并
冪去本自乘爲矩冪減之餘爲實倍高爲法則得折之
高數也
去根如勾折處
如股折梢如弦
通長如股弦和
股弦和與勾求股法曰勾自乘爲實變股弦較乘股弦

In modern algebraic notation we can denote by x the height where the bamboo is broken, and by $10 - x$ the length of the rest of the bamboo. By the Pythagorean theorem,

$$x^2 + 3^2 = (10 - x)^2.$$

Solving this equation gives $x = 4^{11}/_{20}$ chi.

THE *NINE CHAPTERS*

Most ancient Chinese mathematics was written on bamboo or paper which perishes with time. One outstanding survivor, dating possibly from 200BC, is the *Jiu zhang suan shu* (Nine Chapters on the Mathematical Art).

This remarkable work contains 246 questions, with answers but no working shown, and may have been used as a textbook. It deals with both practical and theoretical matters – problems from trade, agriculture, surveying and engineering, as well as discussions of the areas and volumes of various geometrical shapes, the calculation of square roots and cube roots and the study of right-angled triangles. The *Nine Chapters* also includes a discussion of simultaneous equations using a method (now known as *Gaussian elimination*) that was not rediscovered in Europe until 2000 years later.

THE INDIANS

Around 250BC King Asoka, ruler of most of India, became the first Buddhist monarch. His conversion was celebrated around the kingdom by the construction of many pillars carved with his edicts. These Asoka columns included the first known appearance of what would eventually become our *Hindu–Arabic numerals,* a decimal place-value system with separate columns for units, tens, hundreds, etc. From about the year AD400, the Indians also used the number 0, both as a place-holder and as a number to calculate with, and showed how to work with negative numbers.

Indian mathematics can be traced back to a number of Vedic manuscripts from around 600BC. These contain early work on arithmetic, permutations and combinations, the theory of numbers and the extraction of square roots.

Later, in the first millennium AD, the two most outstanding Indian mathematicians were Aryabhata the Elder (b. 476) and Brahmagupta (598–670).

India's *Aryabhata* satellite, 1975

ARYABHATA

One of Aryabhata's main contributions to mathematics was to arithmetic series: these are addition sums like

$$5 + 9 + 13 + 17 + 21 + 25 + 29,$$

where the difference between successive terms is always the same (in this case, 4). Aryabhata gave various rules for adding such numbers together, of which his simplest was:

Add the first and last terms and multiply the answer by half the number of terms.

For the series above, the sum of the first and last terms is $5 + 29 = 34$, and half the number of terms is $3^1/_2$; multiplying these together gives the correct answer, 119.

Aryabhata also presented (in words) formulas for the sum of the first few natural numbers and of their squares and cubes; in our modern notation these are

$$1 + 2 + 3 + \ldots + n = n(n+1)/2$$
$$1^2 + 2^2 + 3^2 + \ldots + n^2 = n(n+1)(2n+1)/6$$
$$1^3 + 2^3 + 3^3 + \ldots + n^3 = n^2(n+1)^2/4$$

— for example (with $n = 10$),

$$1 + 2 + \ldots + 10 = (10 \times 11)/2 = 55$$
$$1^2 + 2^2 + \ldots + 10^2 = (10 \times 11 \times 21)/6 = 385$$
$$1^3 + 2^3 + \ldots + 10^3 = (10^2 \times 11^2)/4 = 3025$$

Aryabhata gave the first systematic treatment of Diophantine equations — algebraic problems for which we seek whole number solutions. He was also interested in trigonometry, constructed tables of the sine function, and obtained the value 3.1416 for π.

The first Indian satellite was named 'Aryabhata' in his honour.

Brahmagupta

BRAHMAGUPTA

Indian mathematicians transformed zero from its role as a place-holder to an actual number to calculate with. In AD628, the astronomer and mathematician Brahmagupta completed a work called the *Brahmasphutasiddhanta* (The Opening of the Universe), in which he began with positive numbers or 'fortunes' (such as 3), explained the use of zero (which he called both *cipher* and *nought*), and then extended his discussion to negative numbers or 'debts' (such as −5) — a great breakthrough. He also gave explicit rules for combining them:

> *The sum of cipher and negative is negative;*
> *of positive and nought, positive;*
> *of two ciphers, cipher.*
> [e.g., $0 + (-5) = -5$, $3 + 0 = 3$, $0 + 0 = 0$]
> *Negative taken from cipher becomes positive,*
> *and positive from cipher is negative;*
> *cipher taken from cipher is nought.*
> [e.g., $0 - (-5) = 5$, $0 - 3 = -3$, $0 - 0 = 0$]
> *The product of cipher and positive,*
> *or of cipher and negative, is nought;*
> *of two ciphers is cipher ...*
> [e.g., $0 \times 3 = 0$, $0 \times (-5) = 0$, $0 \times 0 = 0$]

Brahmagupta also worked extensively on a particular type of Diophantine equation with two unknowns. It is now known as *Pell's equation*, after an incorrect assignation by the 18th-century mathematician Leonhard Euler. This equation has the form

$$Cx^2 + 1 = y^2,$$

and we are required to find whole number solutions for a given value of C. For example, when $C = 3$, we seek whole numbers x and y that satisfy the equation $3x^2 + 1 = y^2$. Two solutions are:

$x = 1$ and $y = 2$, since $(3 \times 1^2) + 1 = 4 = 2^2$;

$x = 4$ and $y = 7$, since $(3 \times 4^2) + 1 = 49 = 7^2$.

But if $x = 2$, then there is no value for y.

Brahmagupta solved Pell's equation for many different values of C, and also derived useful methods for generating new solutions from old; thus, if he could find a single solution to a particular equation, he could then find as many other solutions as he wished.

One particularly difficult case to crack was $61x^2 + 1 = y^2$ (with $C = 61$), where he found the simplest solution

$x = 226{,}153{,}980$ and $y = 1{,}766{,}319{,}049$

— a remarkable achievement. This solution was later rediscovered in the 17th century by the French mathematician Pierre de Fermat.

CYCLIC QUADRILATERALS

One of Brahmagupta's main interests was the study of quadrilaterals whose corners lie on a circle. He obtained formulas for the area of such a quadrilateral and for the lengths of the two diagonals, given the lengths of its four sides, and gave various methods for constructing such quadrilaterals.

THE MAYANS

One of the most interesting counting systems is that of the Mayans of Central America, used between their most productive years from AD300 to 1000. The Mayans were situated over a large area centred on present-day Guatemala and Belize and extending from the Yucatan peninsula of Mexico in the north to Honduras in the south. Most of their calculations involved the construction of calendars, for which they developed a place-value system based mainly on the number 20.

Our knowledge of the Mayan counting system and of their calendars is derived mainly from writings on the walls of caves and ruins, hieroglyphic inscriptions on carved pillars (*stelae*), and a handful of painted manuscripts (*codices*). The codices were intended to guide Mayan priests in ritual ceremonies involving hunting, planting and rainmaking, but many codices were destroyed by the Spanish conquerors who arrived in this area after the year 1500.

The most notable of the surviving codices is the beautiful *Dresden codex,* dating from about 1200. It is painted in colour on a long strip of glazed fig-tree bark and contains many examples of Mayan numbers.

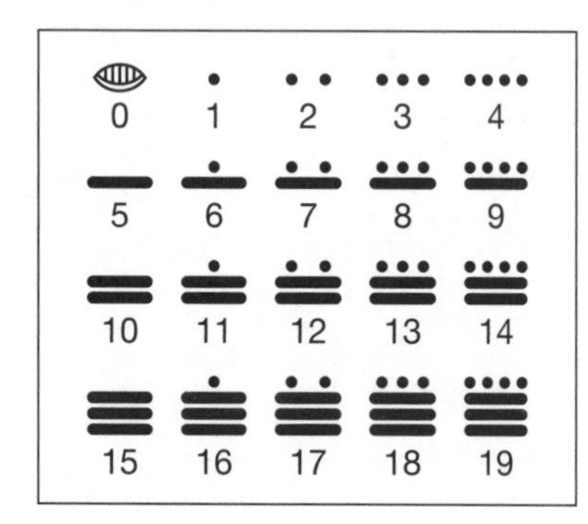

The Mayan symbols for the numbers from 0 to 19.

THE MAYAN NUMBER SYSTEM

The Mayan counting system was a place-value system with a dot to represent 1, a line to represent 5, and a special symbol (a shell) to represent 0. These were combined to give the numbers from 0 to 19.

To obtain larger numbers they combined these numbers, writing them vertically; for example, the illustrated codex depicts the symbol for 12 above the symbol for 13 – this represents the number 253 (12 twenties + 13).

THE HEAD FORM

An interesting feature of the Mayan numerals is that there was an alternative form for each number, a pictorial form or glyph known as the *head-form,* with a pictorial representation of the head of a man, animal, bird or deity. These pictures appear on various pillars: below are the head-forms of various numbers.

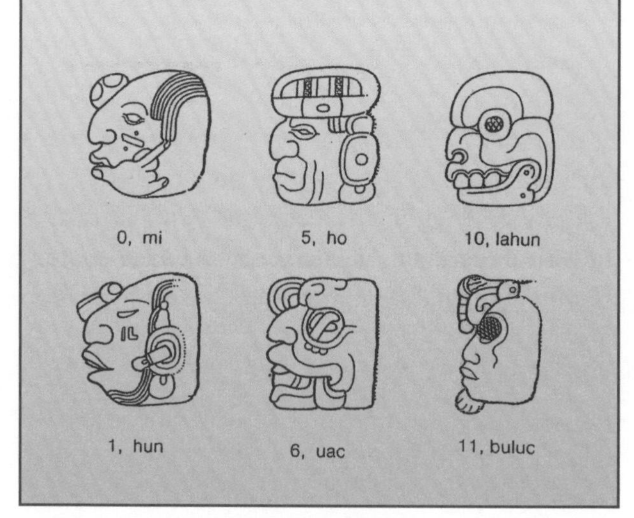

THE MAYAN CALENDARS

In order to keep track of the passage of time, the Mayans employed two types of calendar, with 260 days and 365 days.

Part of a Mayan codex

The 260-day calendar was a ritual one, used for forecasting and known as the *tzolkin*, or 'sacred calendar'. It consisted of thirteen months of twenty days. Each day combined a month-number (from 1 to 13) with one of twenty day-pictures named after deities (such as Imix, Ik and Akbal). These two systems then intermeshed, as illustrated — for example, the day 1 Imix was not followed by 2 Imix and 3 Imix, but by 2 Ik and 3 Akbal, etc. — eventually yielding a cycle of 13 × 20 = 260 days.

For their 365-day calendar, they modified their number system to take account of the number of days in the calendar year. To do so, they introduced an 18 into their 20-based system (since 18 × 20 = 360), and then added five extra 'inauspicious' days to make up the full 365 days. So their counting system was based on the following scheme:

1 kin = 1 day

20 kins = 1 uinal = 20 days

18 uinals = 1 tun = 360 days

20 tuns = 1 katun = 7200 days

20 katuns = 1 baktun = 144,000 days,

and so on. They had no problems in calculating with such large numbers.

These two calendars operated independently, and were also combined to give a *calendar round,* in which the number of days was the least common multiple of 260 and 365, which is 18,980 days, or 52 calendar years. These periods of 52 years were then packaged into even longer time periods. The longest time period used by the Mayans was the *long count* calendar of 5125 years.

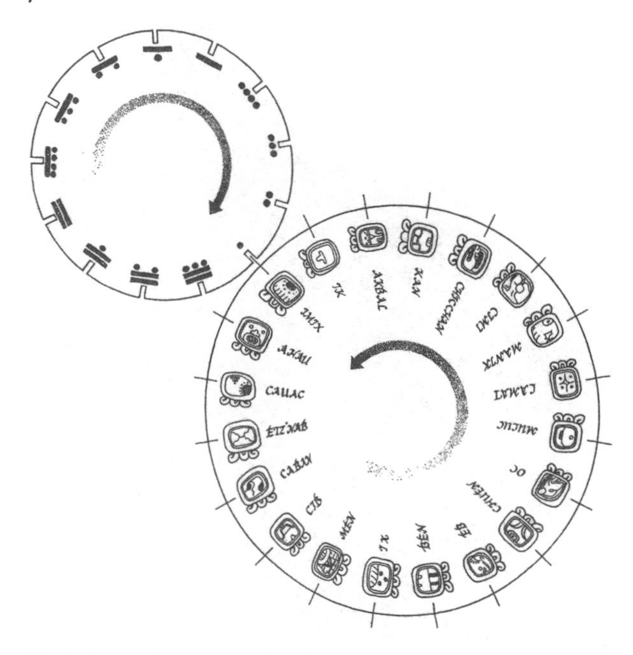

The Mayan 260-day calendar

AL-KHWARIZMI

The period from 750 to 1400 saw an awakening of interest in Greek and Indian culture in Mesopotamia. Inspired by the teachings of the prophet Muhammad, Islamic scholars seized on the ancient texts, translating them into Arabic and extending and commenting on them. Baghdad, on the trade routes for silk and spices, was well placed to receive the writings of Greek geometers and the contributions of Indian scholars — including the positional method of counting.

Early Islamic astronomers use a theodolite

In Baghdad the caliphs actively promoted mathematics and astronomy, and in the early 9th century Caliph Harun al-Rashid and his son al-Ma'mun established and supported the 'House of Wisdom', a scientific academy with its own extensive library and observatory. There, Islamic mathematicians translated and commented on the Greek works of Euclid, Archimedes and others, and developed the Indian decimal place-value counting system into what are now the *Hindu–Arabic numerals.*

AL-KHWARIZMI (c.783–c.850)

One of the earliest scholars at the House of Wisdom was the Persian scholar Muhammad ibn-Musa (al-)Khwarizmi (his Persian name omitted the Arabic prefix 'al-'). The author of two celebrated astronomical star tables and an influential treatise on the astrolabe, he is

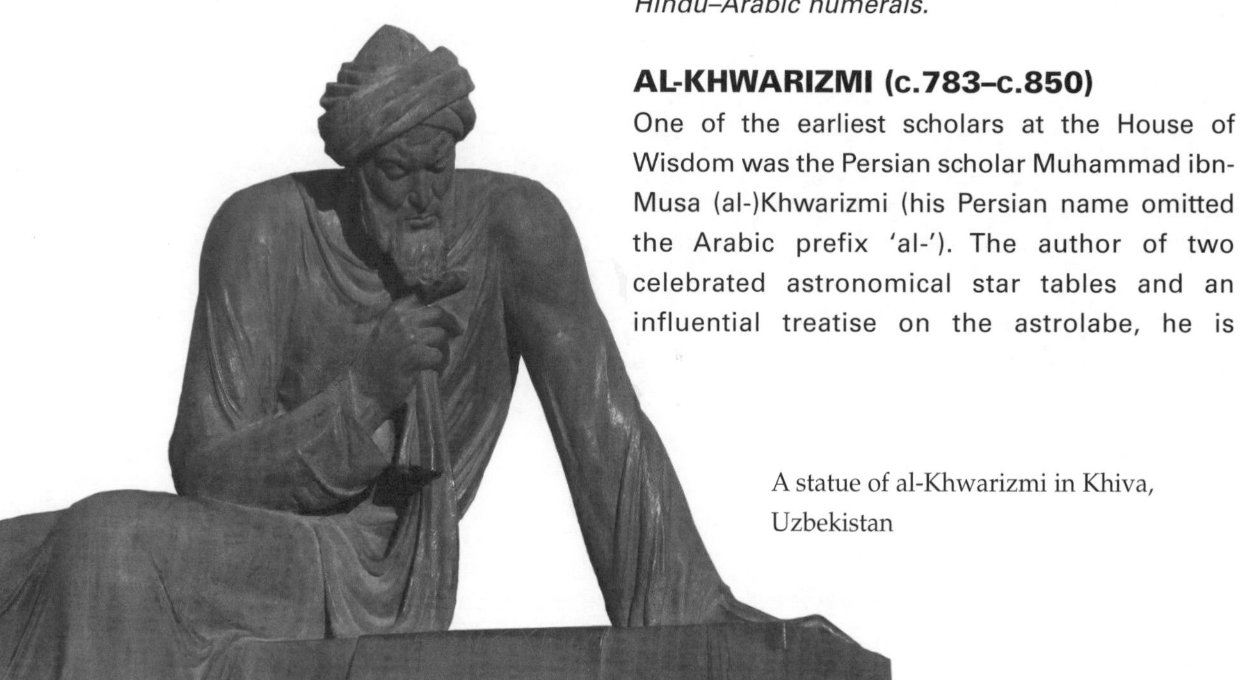

A statue of al-Khwarizmi in Khiva, Uzbekistan

SOLVING A QUADRATIC EQUATION

One square, and ten roots of the same, amount to thirty-nine dirhems
(a dirhem is a unit of currency)

In modern notation, this is $x^2 + 10x = 39$.

To solve this, al-Khwarizmi started with a square of side x (shaded) and added two rectangles of length x and width 5 (note that 5 is half of 10). He then completed the square by adding a new square of side 5.

The resulting larger square of side $x + 5$ has area $(x + 5)^2$, and is made up from two smaller squares with areas x^2 and 25 and two rectangles each with area $5x$. Thus,

$$(x + 5)^2 = x^2 + 10x + 25.$$

Since $x^2 + 10x = 39$, we have

$$(x + 5)^2 = 39 + 25 = 64.$$

Taking the square root, he found that $x + 5 = 8$, giving the solution $x = 3$.

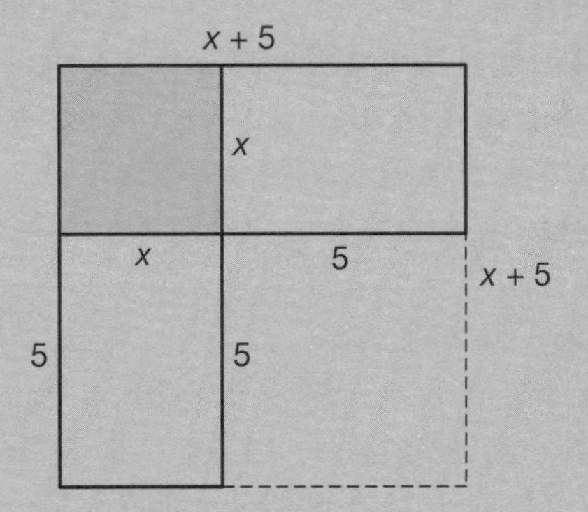

Note that the other solution (−13), being a negative number, was not considered meaningful and was ignored.

remembered by mathematicians primarily for his books on arithmetic and algebra.

Neither book contained results of great originality, but his *Arithmetic* was important for introducing the Indian number system to the Islamic world and later helping to spread the decimal counting system throughout Christian Europe. Indeed, his Arabic name, transmuted into 'algorism', was later used in Europe to mean arithmetic, and we still use the word *algorithm* to refer to a step-by-step procedure for solving problems.

The title of al-Khwarizmi's algebra book is *Kitab al-jabr wal-muqabala* (The Compendious Book on Calculation by Completion [*al-jabr*] and Reduction [*al-muqabala*]). This book title is the origin of our word 'algebra': the term 'al-jabr' refers to the operation of transposing a term from one side of an algebraic equation to the other.

Al-Khwarizmi's *Algebra* commences with a lengthy account of how to solve linear equations (with numbers and terms involving x) and quadratic equations (also involving x^2). Since negative numbers were still not considered meaningful, he split the equations into the following six types, given here with their modern equivalents (where a, b and c are positive constants):

roots equal to numbers ($ax = b$)
squares equal to numbers ($ax^2 = b$)
squares equal to roots ($ax^2 = bx$)
squares and roots equal to numbers ($ax^2 + bx = c$)
squares and numbers equal to roots ($ax^2 + c = bx$)
roots and numbers equal to squares ($bx + c = ax^2$)

He then proceeded to solve instances of each type, such as $x^2 + 10x = 39$ (see above), using a geometrical form of 'completing the square'.

ALHAZEN AND OMAR KHAYYAM

Influential among Arabic geometers was ibn al-Haitham (965–1039). Known in the west as Alhazen, his main contributions were to the study of optics. The Persian poet Omar Khayyam (1048–1131), remembered mainly for his collection of poems called the *Rubaiyat*, was also a mathematician who wrote on algebra, geometry and the calendar.

Alhazen's widespread contributions to optics include his inventions of the pinhole camera and the camera obscura. In Book V of his influential seven-volume *Kitab al-Manazir* (Book of Optics), he proposed and answered a question now celebrated as 'Alhazen's problem':

At which point on a spherical mirror is light from a given point source reflected into the eye of a given observer?

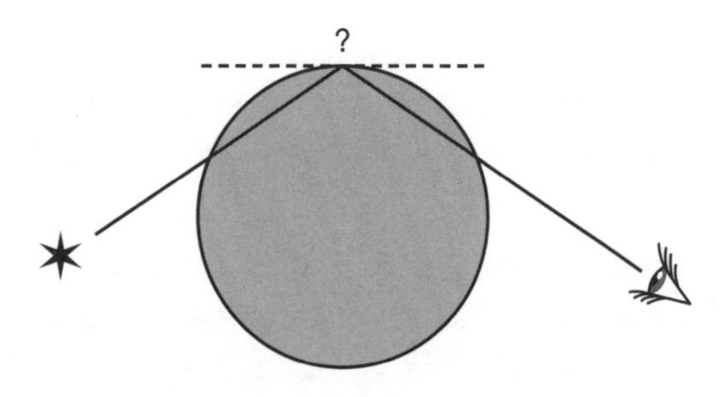

Alhazen's problem

This problem asks for the point on the circumference of a given circle at which lines from two given points meet and make equal angles with the tangent there. A related problem is:

At which point on the cushion of a circular billiard table must a cue ball be aimed so as to hit a given target ball?

THE PARALLEL POSTULATE

Alhazen also tried to prove Euclid's 'parallel postulate', which we now describe.

Euclid's *Elements* opens with five postulates that we assume to be true: as we saw, these include the observation that we can draw a circle with any given centre and radius. But while the first four postulates are short and simple, the fifth is more complicated:

If two lines include angles x and y whose sum is less than 180°, *then these lines must meet if extended indefinitely.*

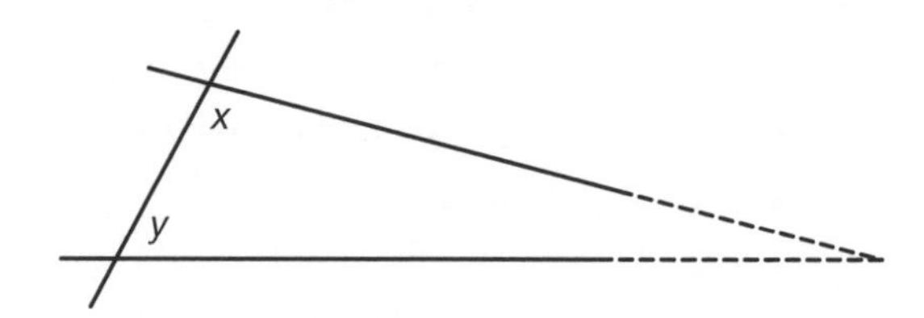

This postulate seems like a result that one ought to be able to prove, rather than an assumed fact, and for two thousand years generations of geometers tried to deduce it from the other four postulates.

One approach to proving it was to find another result 'equivalent' to it: if we can then prove this other result, then the fifth postulate follows. One such equivalent result is:

Given any line L and any point P that does not lie on this line, there is exactly one line, parallel to L, that passes through P.

This equivalent version is illustrated below: because of it, Euclid's fifth postulate is usually called the *parallel postulate.*

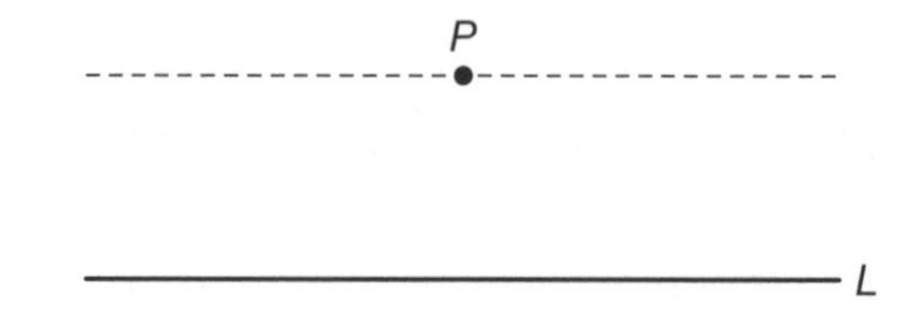

The problem of the parallel postulate remained unresolved until the 19th century.

Alhazen was one of the first to try to prove the parallel postulate. His method was ingenious. He dropped a perpendicular from the point *P* to the line *L*, and then moved this perpendicular line to left and right, as shown; its top end then traced out the required line through *P* parallel to the line *L*.

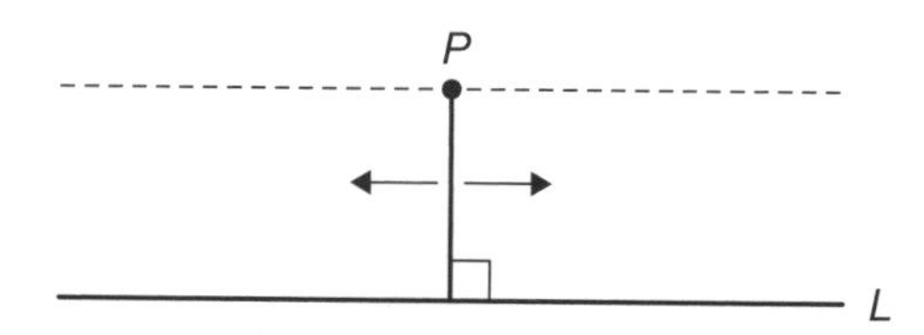

Alhazen's 'proof' of the parallel postulate

OMAR KHAYYAM

One person who publicly criticized Alhazen's attempt to prove the parallel postulate was Omar Khayyam (in Arabic, al-Khayyami), who also wrote about algebra and mechanics. His refutation was on the grounds that using motion to prove results is not permitted in the context of Euclid's *Elements*:

> *There are many things wrong here.*
> *How could a line move, remaining perpendicular to a given line?*
> *How could a proof be based on this idea?*
> *How could geometry and motion be connected?*

Omar Khayyam

In his writings on algebra Omar Khayyam presented the first systematic classification of cubic equations (those involving x^3), similar to al-Khwarizmi's classification of linear and quadratic equations. He also presented a geometrical method for solving several types of cubic equations; for example, to solve

A solid cube plus edges equal to a number

(in modern notation, $x^3 + cx = d$), his method was to draw a particular semicircle ($x^2 + y^2 = (d/c)x$) and a particular parabola ($x^2 = \sqrt{c}\, y$) and find the point where they intersect: a solution x is as marked.

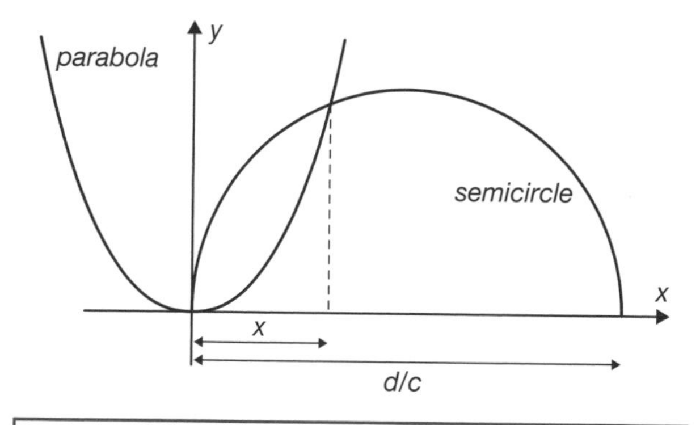

REFORMING THE CALENDAR

Another interest of Omar Khayyam was calendar reform. He was one of a group of eight learned men at the imperial observatory in Isfahan who were commissioned by Sultan Malik Shah I to replace the Persian lunar calendar by a solar calendar. The result, the Jalali calendar, lasted for eight centuries and was very accurate – in particular, their estimate of 365.24219858156 days for the length of a year was only a few seconds out.

CHAPTER 2

EARLY EUROPEAN MATHEMATICS

The revival of mathematical learning during the Middle Ages was largely due to three factors:

- **the translation of Arabic classical texts into Latin during the 12th and 13th centuries**
- **the establishment of the earliest European universities**
- **the invention of printing**

The first of these made the works of Euclid, Archimedes and other Greek writers available to European scholars, the second enabled groups of like-minded scholars to meet and discourse on matters of common interest, while the last enabled scholarly works to be available at modest cost to the general populace in their own language.

The first European university was founded in Bologna in 1088, and Paris and Oxford followed shortly after. The curriculum was in two parts. The first part, studied for four years by those aspiring to a Bachelor's degree, was based on the ancient 'trivium' of grammar, rhetoric and logic (usually Aristotelian). The second part, leading to a Master's degree, was based on the 'quadrivium', the Greek mathematical arts of arithmetic, geometry, astronomy and music; the works studied included Euclid's *Elements* and Ptolemy's *Almagest.*

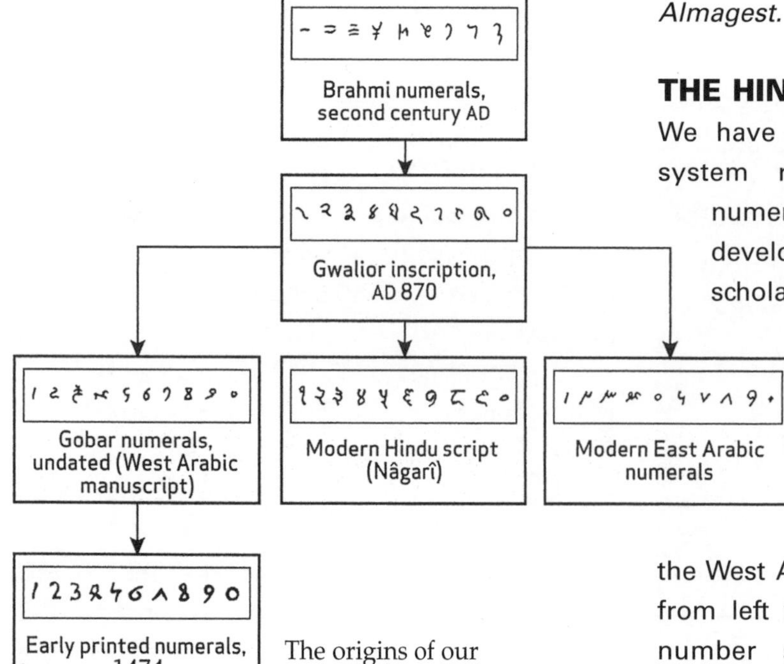

The origins of our number systems

THE HINDU–ARABIC NUMERALS

We have seen how the decimal place-value system represented by the Hindu–Arabic numerals first arose in India and was later developed by al-Khwarizmi and other Islamic scholars working in Baghdad and elsewhere. Gradually the numerals diverged into three separate types — the modern Hindu script, the East Arabic numerals (written from right to left), still found today in the countries of the Middle East, and the West Arabic numerals 1 to 9 and 0 (written from left to right) that eventually became the number system used throughout Western Europe.

Viewing the heavens with a joynt rule

But it took many centuries for the Western form of the Hindu–Arabic numerals to become fully established. They were certainly more convenient to calculate with than Roman numerals, but for practical use most people continued to use an abacus.

As time progressed the situation improved with the publication of influential books that promoted them, such as those by Fibonacci (in Latin), Pacioli (in Italian) and Recorde (in English). By the time that printed books had become widely available, the Hindu–Arabic numerals were in general use.

THE AGE OF DISCOVERY

The spirit of enquiry and inventiveness of the Middle Ages and the Renaissance led people to adopt a more critical view to ideas that had been accepted for centuries. It showed itself in many ways:

- the voyages of discovery to unknown lands
- the development and invention of scientific and mathematical instruments for a variety of purposes
- the use of geometrical perspective in painting and other visual arts
- the solution of cubic and quartic equations
- the development and standardization of mathematical terminology and notation
- the revolutionary approach to planetary motion
- the rediscovery and reinterpretation of classical texts
- the development of mechanics
- the removal of algebra from its dependence on geometry.

These all contributed to the development of a view that the universe is a book written in the language of mathematics. As instruments became ever more sophisticated, mathematics for practical purposes increased – particularly in navigation, map-making, astronomy and warfare.

GERBERT

The period from 500 to 1000 in Europe is known as the Dark Ages. The legacy of the ancient world was almost forgotten, schooling became infrequent, and the general level of culture remained low. Apart from a few sporadic writings (by the Venerable Bede, Alcuin of York, and others) on the calendar, finger reckoning and arithmetical problems, mathematical activity was generally sparse.

During the 8th and 9th centuries, the Islamic world spread along the northern coast of Africa and up through southern Spain and Italy, and Moslem schools were established in Catalonia, while Córdoba became the scientific capital of Europe.

Islamic decorative art and architecture also spread throughout southern Spain: celebrated examples include the magnificent geometrical arches in the Córdoba Mezquita (mosque) and the variety of geometrical tiling patterns in the Alcázar in Seville and the Alhambra in Granada.

An Arabic tiling in the Seville Alcázar

The mosque at Córdoba

GERBERT OF AURILLAC

Revival of interest in mathematics is generally believed to begin with Gerbert of Aurillac (c.940–1003). Gerbert entered the Church and was trained in Aurillac in southern France. Because of his interest in the sciences he was sent to study in Catalonia, and later to Córdoba and Seville, where he encountered the achievements of the Islamic world, such as the development of the Hindu–Arabic numerals and the use of the astrolabe.

Gerbert travelled widely, becoming renowned as a teacher of the quadrivium subjects (arithmetic, geometry, astronomy and music). He is believed to have been the first to introduce the Hindu–Arabic numerals to Christian Europe, using an abacus that he designed especially for the purpose.

Gerbert also reintroduced the *armillary sphere,* an astronomical instrument invented by the Greeks. Built from interlocking metal circular rings, it illustrated the motions of the sun, the planets and the stars as they were perceived to travel around the earth.

THE ASTROLABE

An *astrolabe* is an instrument that was widely used by astronomers, navigators, surveyors and religious leaders in the Islamic world and the Europe of the Middle Ages. Once fully developed, it could be used for finding and predicting the positions of the sun, moon and stars, for calculating latitude, for determining the direction of Mecca and times of prayer, for carrying out calculations, and for the casting of horoscopes.

The invention of the astrolabe has been traced back to Ancient Greece – in particular, both Hipparchus and Ptolemy have been credited with its use. Theon of Alexandria (Hypatia's father) wrote a treatise on the astrolabe, and Hypatia gave lectures about it.

An astrolabe

During the Islamic age (first in Mesopotamia and then in southern Europe) the instrument came into its own. Angular and numerical scales appeared around the rim and a sighting bar for use in astronomy was added. In later years a more convenient version, the *mariners' astrolabe,* was developed for use at sea.

In 995 Gerbert became tutor to Emperor Otto III in the imperial court at Rome. Throughout this time Gerbert had risen steadily in the Church, and was appointed Archbishop of Ravenna in 998. Through the Emperor's influence he was crowned Pope Sylvester II in 999.

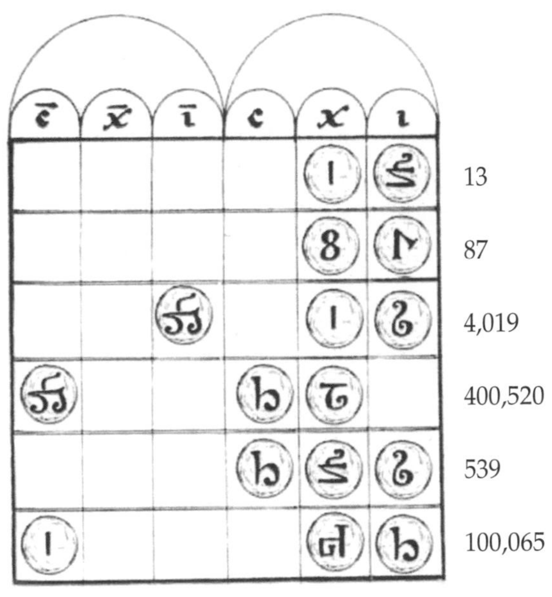

An abacus of the type used by Gerbert and his followers

Gerbert's statue in Aurillac

FIBONACCI

Leonardo of Pisa (c.1170–1240), known since the 19th century as Fibonacci (son of Bonaccio), is remembered mainly for his *Liber Abaci* (Book of Calculation), which he used to popularize the Hindu–Arabic numerals, and for a number sequence named after him. His work was crucial in bringing Arabic mathematics to wider recognition in Western Europe.

Fibonacci was born in Pisa. After travelling widely throughout the Mediterranean, he returned home and wrote works expanding on what he had learned, to help his countrymen deal with calculation and commerce.

THE *LIBER ABACI*

Most of our knowledge about Fibonacci comes from the prologue of his influential book *Liber Abaci.* The first edition of this book appeared in 1202. It covers four main areas starting with the use of Hindu–Arabic numerals in calculation and then using them for the mathematics needed in business. The largest part of the book deals with recreational mathematical problems, finishing with operations on roots and a little geometry.

Leonardo Fibonacci

PROBLEMS FROM THE *LIBER ABACI*

Fibonacci's *Liber Abaci* contains a wide range of mathematical problems, including the following three that may be similar to ones you remember from your school days!

> *There is a tree,* $\frac{1}{4}$ *and* $\frac{1}{3}$ *of which lie below ground. If the part below ground is* 21 *palmi, how tall is the tree?*
>
> *If a lion can eat a sheep in* 4 *hours, a leopard can eat it in* 5 *hours, and a bear can eat it in* 6 *hours, how long would they take eating it together?*
>
> *I can buy* 3 *sparrows for a penny,* 2 *turtle-doves for a penny or doves for* 2 *pence each. If I spent* 30 *pence buying* 30 *birds and bought at least one bird of each kind, how many of each kind did I buy?*

Another problem involves adding powers of 7:

> *7 old women are going to Rome;*
> *each has 7 mules; each mule carries 7 sacks;*
> *each sack contains 7 loaves; each loaf has*
> *7 knives; each knife has 7 sheaths;*
> *what is the total number of things?*

This is reminiscent of a problem from the Egyptian Rhind papyrus:

> *houses* 7; *cats* 49; *mice* 343; *spelt* 2401;
> *hekat* 16,807. *Total* 19,607

and also of the more recent nursery rhyme:

> *As I was going to St Ives I met a man with 7 wives ... Kits, cats, sacks and wives,*
> *How many were going to St Ives?*

Such examples dramatically illustrate the fact that the same mathematical idea can resurface in different guises over thousands of years.

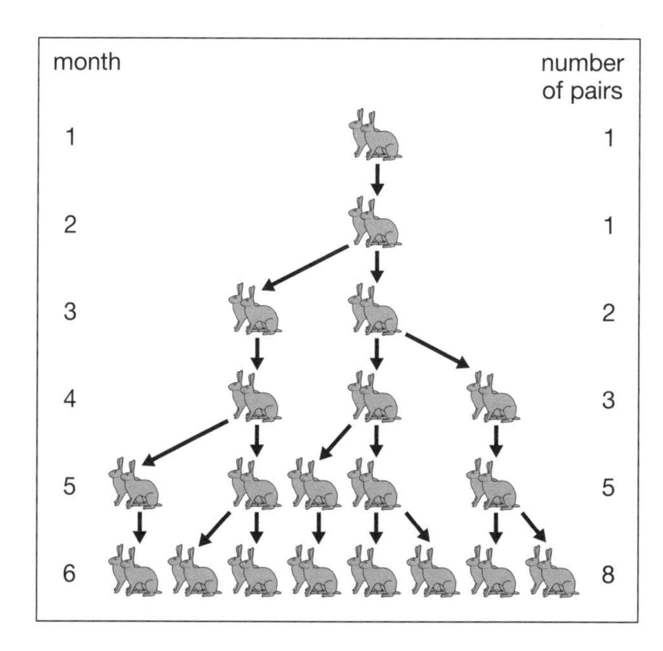

THE RABBITS PROBLEM

The most famous problem in the *Liber Abaci* is the problem of the rabbits:

> *A farmer has a pair of baby rabbits. Rabbits take two months to reach maturity and then give birth to another pair each month. How many pairs of rabbits are there after a year?*

To solve this, we note that:

- In months 1 and 2 the farmer has only the original pair,
- In month 3, a new pair arrives, so he now has two pairs,
- In month 4, the original pair produces another pair, and the new pair has not yet produced, so he now has three pairs,
- In month 5, the original pair and the new pair both produce another pair; and so on.

The result of the problem is that the number of rabbits in each month follows the so-called *Fibonacci sequence:*

1, 1, 2, 3, 5, 8, 13, 21, 34, 55, 89, 144, ...,

in which each successive number (after the first two) is the sum of the previous two; for example, 89 = 34 + 55. The answer to the problem is the 12th number, which is 144.

SPIRALS AND THE GOLDEN NUMBER

The ratios of successive terms of the Fibonacci sequence are

$1/1, 2/1, 3/2, 5/3, 8/5, \ldots$.

These tend to the 'golden number'

$\varphi = \frac{1}{2}(1 + \sqrt{5}) = 1.618...,$

which has remarkable and pleasing properties: for example, to find its square we add 1 ($\varphi^2 = 2.618...$), and to find its reciprocal we subtract 1 ($1/\varphi = 0.618...$).

A rectangle whose sides are in the ratio φ to 1 is often considered to have the most pleasing shape — neither too thin, nor too fat. The following picture shows how the Fibonacci numbers can be arranged so as to give rise to a spiral pattern; further rectangles can be added at will.

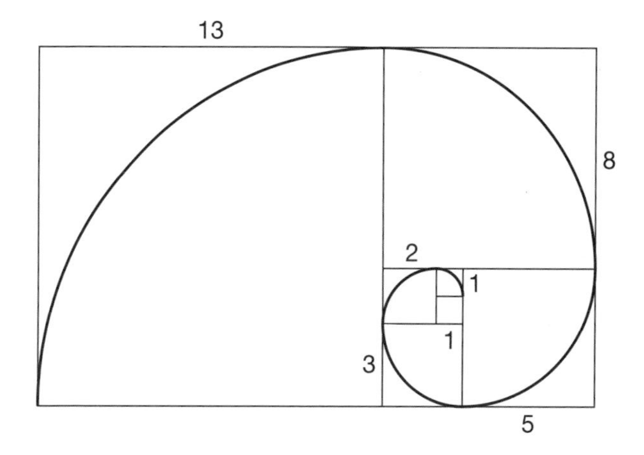

Similar spirals occur through nature — on a nautilus shell and in the pattern of seeds in a sunflower — for example, the number of seeds in such a spiral pattern is often 34, 55 or 89, all of which are Fibonacci numbers.

EARLY OXFORD MATHEMATICIANS

The actual foundation date of Oxford University is uncertain, but by the beginning of the 13th century the University had a recognized head who in 1214 was given the official title of 'Chancellor'. This was Bishop Grosseteste (c.1175–1253) who founded the tradition of scientific thought in Oxford.

Grosseteste was particularly interested in geometry and optics, and wrote in praise of mathematics as follows:

> *The usefulness of considering lines, angles and figures is the greatest, because it is impossible to understand natural philosophy without them ... By the power of geometry, the careful observer of natural things can give the causes of all natural effects ...*

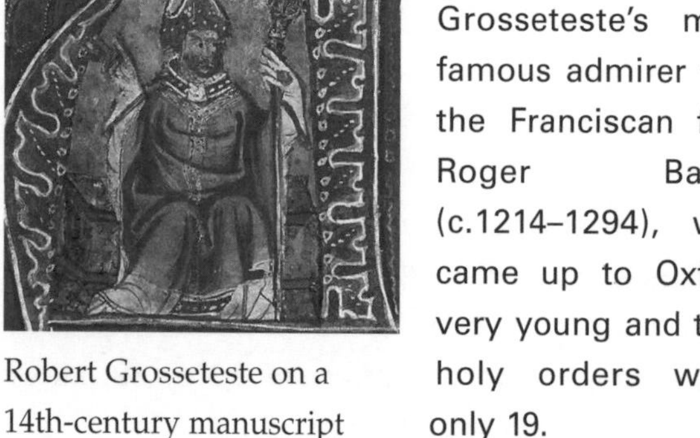

Robert Grosseteste on a 14th-century manuscript

Grosseteste's most famous admirer was the Franciscan friar Roger Bacon (c.1214–1294), who came up to Oxford very young and took holy orders when only 19.

Known as 'Dr Mirabilis', Bacon spent most of his money on scientific manuscripts and instruments and wrote on scientific issues – but this led him into conflict with the Church in Rome and he was imprisoned for his views. Like Grosseteste, he believed that:

> *He who knows not mathematics cannot know the other sciences nor the things of this world. And those who have no knowledge of mathematics do not perceive their own ignorance and so do not look for a cure.*

Roger Bacon

Bacon's study was on Folly Bridge over the River Thames – an observatory that soon became a place of pilgrimage for scientists. As Samuel Pepys wrote in his diary in 1669:

> *So to Friar Bacon's study: I up and saw it and gave the man a shilling. Oxford mighty fine place.*

Friar Bacon's study

GEOFFREY CHAUCER (1342–1400)

Chaucer is mainly remembered as the author of *The Canterbury Tales.* In one tale Nicholas, a poor Oxford scholar, kept by his bed a copy of Ptolemy's *Almagest,* an astrolabe and 'augrim-stones' (counters for calculations):

His Almageste and bokes grete and smale,
His astrelabie, longinge for his art,
His augrim-stones layen faire a-part
On shelves couched at his beddes heed.

Chaucer was interested in mathematical instruments, and his *Treatise on the Astrolabe* (1393) was one of the earliest science books to appear in English.

The Chaucer astrolabe

THE MERTON SCHOOL

By the beginning of the 14th century, scholars had started to organize themselves into colleges, and three colleges were in existence. Merton College quickly became pre-eminent in scientific studies, and the Merton School became famous throughout Europe. Members of this School tried to treat mathematically all sorts of natural phenomena, such as heat, light, forces, density and colour, and even tried to quantify knowledge, grace and charity.

Richard of Wallingford measuring a disc with a pair of compasses

Associated with the Merton School was Richard of Wallingford (1292–1336), who studied at Oxford before becoming abbot of St Albans. He wrote the earliest Latin treatise on trigonometry and devised and built mathematical instruments for use in astronomy and navigation. He is best known for his astronomical clock (now in St Albans Cathedral), which he described in his 1327 *Tractatus Horologii Astronomici* (Treatise on the Astronomical Clock).

The most important of the Merton Scholars was Thomas Bradwardine (c.1290–1349), the greatest English mathematician of the 14th century. He wrote influential books on topics ranging from arithmetic and algebra to velocities and logic, and his discourses were so learned that he was called 'Dr Profundus'. Elevated to the position of Archbishop of Canterbury in 1349, he died of the Black Death a few weeks later.

About mathematics he wrote:

Mathematics reveals every genuine truth, for it knows every hidden secret, and bears the key to every subtlety of letters. Whoever then has the effrontery to study physics while neglecting mathematics should know from the start that he will never make his entry through the portals of wisdom.

ORESME

Nicole Oresme (c.1323–1382) was born near Caen in Normandy, and studied at the University of Paris, receiving a doctorate in theology. He was subsequently Grand Master of the College of Navarre, Dean of Rouen Cathedral, and Bishop of Lisieux. A friend of King Charles V, Oresme carried out translations of the works of Aristotle at the king's request. In mathematics he studied infinite series and proportion and anticipated later work on mechanics and the representation of data in graphical form.

Oresme was opposed to many of Aristotle's ideas, especially on weight and planetary motion. In his *Traité du Ciel et du Monde* (Treatise on the Heaven and the Earth) he considered various arguments for and against the idea that the earth rotates on its axis, explained why Aristotle's arguments for a static earth were not valid, and established that no experiment could determine whether the earth travels from west to east or the heavens move from east to west.

Nicole Oresme

Oresme also believed that if the earth were moving, rather than the celestial spheres that carried the stars and planets with them, then any observations made from the earth would remain unchanged. But in spite of anticipating several of Copernicus's conclusions, he considered his own arguments to be merely speculative:

> *Everyone maintains, and I think myself, that the heavens do move and not the earth.*

GRAPHICAL REPRESENTATION

In his *Tractatus de Configurationibus Qualitatum et Motuum* (Treatise on Configurations of Qualities and Motion) Oresme studied the nature of heat along a rod and other 'qualities' such as whiteness and sweetness. He distinguished between the intensity of heat at each point (the *intensio* or *latitude*) and the length of the heated rod (the *extensio* or *longitude*).

He then depicted these two quantities in a diagram, rather like our two-dimensional rectangular coordinates, with the longitudes along a base line (the horizontal axis) and the latitude at each point of the rod represented by a vertical line of the appropriate height at the point in question.

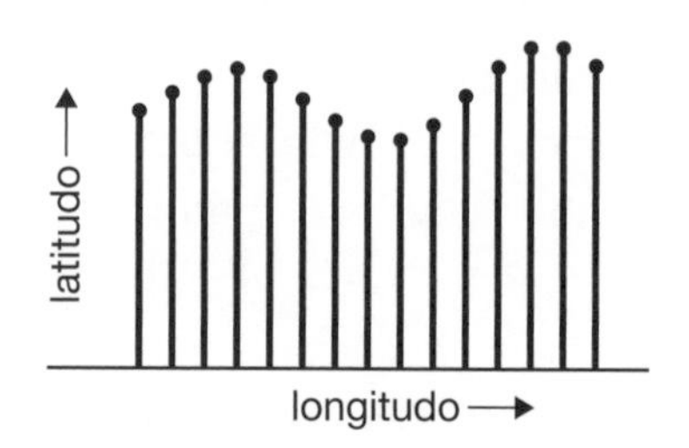

Oresme's graphical representation

INFINITE SERIES

Oresme contributed to several areas of mathematics. Here we present his celebrated proof that the 'harmonic series' $1 + \frac{1}{2} + \frac{1}{3} + \frac{1}{4} + \frac{1}{5} + \ldots$ does not converge to any finite sum, but increases without limit.

To this end, he first grouped the terms:

$$1 + \tfrac{1}{2} + (\tfrac{1}{3} + \tfrac{1}{4}) + (\tfrac{1}{5} + \tfrac{1}{6} + \tfrac{1}{7} + \tfrac{1}{8}) + \ldots .$$

He then noted that this sum is larger than

$$1 + \tfrac{1}{2} + (\tfrac{1}{4} + \tfrac{1}{4}) + (\tfrac{1}{8} + \tfrac{1}{8} + \tfrac{1}{8} + \tfrac{1}{8}) + \ldots = 1 + \tfrac{1}{2} + \tfrac{1}{2} + \tfrac{1}{2} + \ldots ,$$

since the numbers in each bracket add up to $\frac{1}{2}$.

But this last series has no finite sum, and therefore nor does the original one.

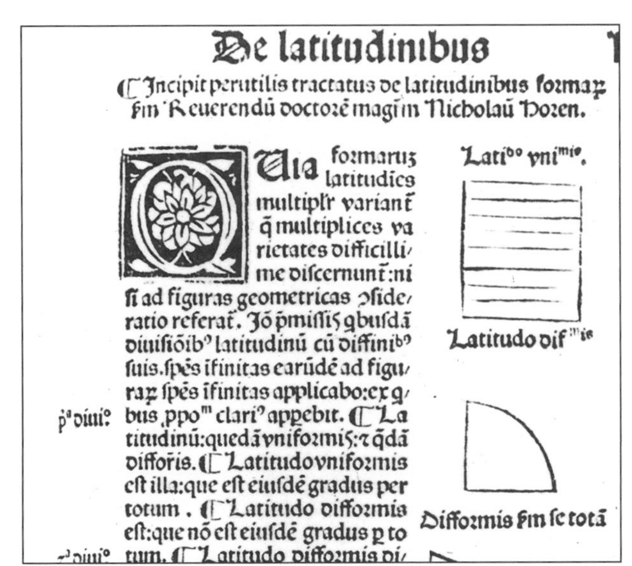

De latitudinibus

¶ Incipit perutilis tractatus de latitudinibus formaꝝ ſm Reuerendū doctorē magiſtrū Nicholaū Horen.

Quia formarū latitudines multiplr variantur q̄ multiplices varietates difficillime diſcernunt: niſi ad figuras geometricas ꝯſideratio referat. Jdō pmiſſis qbuſdā diuiſioib⁹ latitudinū cū diffinitib⁹ ſuis. ſpēs īfinitas earūdē ad figuraꝝ ſpēs īfinitas applicabo: ex qbus ppo^m clari⁹ appebit. ¶ Latitudinū: quedā vniformis: ⁊ qdā difformis. ¶ Latitudo vniformis eſt illa: que eſt eiuſdē gradus per totum. ¶ Latitudo difformis eſt: que nō eſt eiuſdē gradus p totum. ¶ Latitudo difformis di…

p^a diuiſ^o

Lati^do vni^mis.

Latitudo dif^mis

Difformis ſm ſe totā

An extract from Oresme's *Latitude of Forms*

APPLICATION TO MOTION

Oresme also carried out a similar analysis for the motion of an object. Here the longitude is the time taken and the latitude is the speed of the object. He then anticipated Galileo by noting that the area traced out by the lines of latitude corresponds to the distance travelled after a given time.

Another of Oresme's results, also later credited to Galileo, involved average speeds:

> *The distance travelled in a fixed time by a body moving under uniform acceleration is the same as if the body moved at a uniform speed equal to its speed at the midpoint of the time period.*

The resulting picture (or *configuration*) was then used to describe the behaviour of the quality. A horizontal line corresponded to *uniform behaviour* (such as constant heat intensity), while a straight line emerging at an angle to the base line represented *uniformly difform* behaviour (with the latitude increasing steadily as one progressed along the rod).

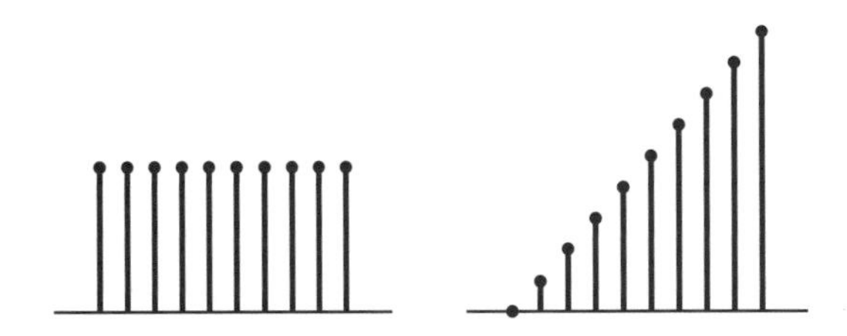

Uniform and uniformly difform behaviour

Oresme proved this by observing that the area of the triangle *ABC* below is the same as the area of the rectangle *ABGE*.

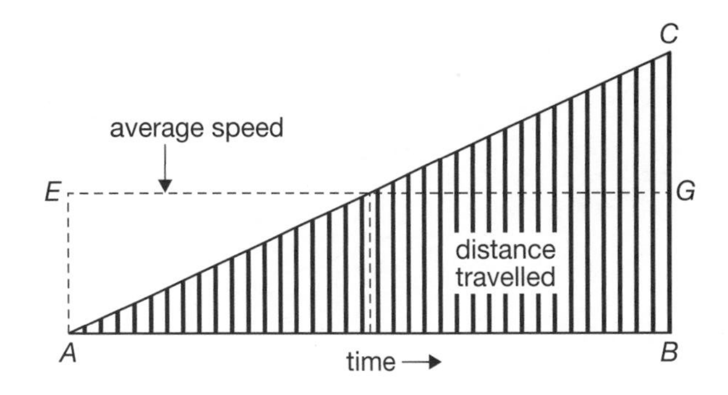

Oresme's result on average speeds

REGIOMONTANUS

Johannes Müller von Königsberg (1436–1476), called Regiomontanus, was probably the most important astronomer of the 15th century. He analysed the discrepancies between observation and theory, and it is often claimed that he set the agenda for the reform of astronomy to which Copernicus, Tycho Brahe and Kepler later contributed. His two most influential works were his writings on Ptolemy's *Almagest* and on triangles, both published after his death.

Regiomontanus with an astrolabe

Epitome of Ptolemy's Almagest

Regiomontanus was born near the town of Königsberg in Bavaria (not the more famous Königsberg in East Prussia). He studied at the Universities of Leipzig and Vienna and was appointed to a position at the latter university in 1457.

EPITOME OF PTOLEMY'S ALMAGEST

His collaborator at Vienna, Georg Peuerbach, died in 1461, requesting on his deathbed that Regiomontanus should continue his efforts in summarizing and commenting on Ptolemy's *Almagest.* The work was published posthumously as *Epytoma in Almagesti Ptolemei* (Epitome of Ptolemy's *Almagest*) in 1496. Giving western Europeans their first accessible and authoritative account of Ptolemy's astronomy, it was studied by every astronomer of note in the 16th century. Its frontispiece shows Ptolemy and Regiomontanus sitting beneath a large armillary sphere.

Regiomontanus returned to Nurnberg (near his birthplace) in 1471, founded his own printing press, and was one of the first publishers of mathematical and scientific work for commercial use.

In 1474 Regiomontanus published tables giving the positions of the sun, moon and planets for the next thirty years — tables that Christopher Columbus used to predict the lunar eclipse of 29

REGIOMONTANUS'S MAXIMUM PROBLEM

In 1471 Regiomontanus asked:

From which point P on the ground does a perpendicularly suspended rod AB appear largest **(and so make the angle θ greatest)?**

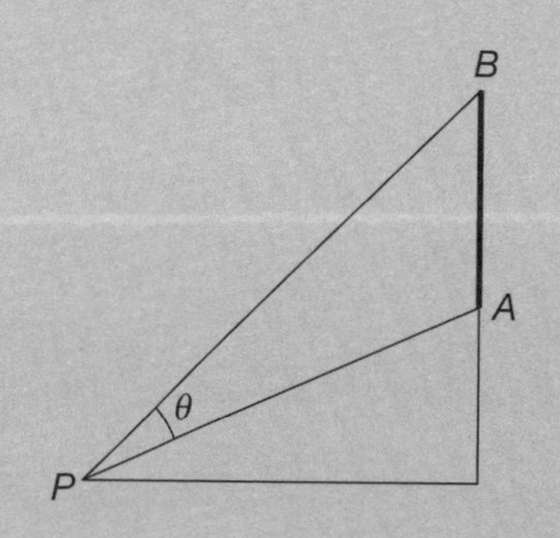

We do not know what caused him to pose this problem. It may have arisen from the new subject of perspective, perhaps to give the best position from which to view the window of a building. It is thought to be one of the first 'extremal' problems since antiquity.

February 1504 during his fourth voyage to the New World; he used this prediction to his advantage to impress and intimidate the native population.

TRIGONOMETRY

Because they needed to calculate the relationships between the angles and sides of various figures, many astronomers became involved with the development of trigonometry. Hipparchus, the Father of Trigonometry, was an astronomer, as was Ptolemy, while Islamic and Hindu astronomers expanded on the Greek tradition, especially in spherical trigonometry.

In *De Triangulis Omnimodis* (On Triangles of Every Kind) Regiomontanus continued this development, systematically organizing his previous trigonometric work and modelling his approach on that of Euclid's *Elements*.

DOCTISSIMI VIRI ET MATHE-
maticarum disciplinarum eximij professoris

IOANNIS DE RE-
GIO MONTE DE TRIANGVLIS OMNI-
MODIS LIBRI QVINQVE:

Quibus explicantur res necessariæ cognitu, uolentibus ad scientiarum Astronomicarum perfectionem deueni-re:quæ cum nusquã alibi hoc tempore expositæ habeantur, frustra sine harum instructione ad illam quisquam aspirarit.

Accesserunt huc in calce pleraq; D. Nicolai Cusani de Quadratura circuli, Deq; recti ac curui commensuratione: itemq; Io. de monte Regio eadem de re ἐλεγκτικά, hactenus à nemine publicata.

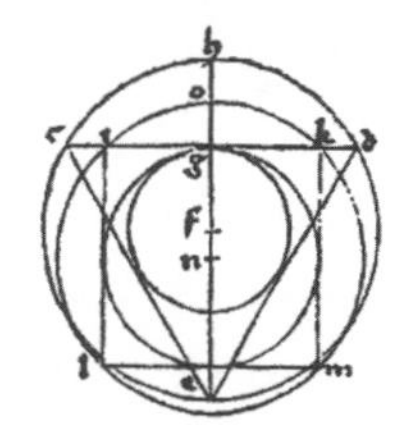

Omnia recens in lucem edita, fide & diligentia singulari. Norimbergæ in ædibus Io. Petrei.

ANNO CHRISTI

M. D. XXXIII.

De Triangulis Omnimodis

It consists of five books. The first contains definitions and axioms, followed by geometrical solutions of plane triangles. The trigonometry starts in the second book, where we see for the first time a result that implies the formula for the area of a triangle in terms of the lengths of two sides and the angle between them. The last three books deal with spherical geometry and trigonometry.

His aim was to provide a mathematical introduction to astronomy, and he wrote:

You, who wish to study great and wonderous things, who wonder about the movement of the stars, must read these theorems about Triangles ... For no one can bypass the science of triangles and reach a satisfying knowledge of the stars ... A new student should neither be frightened nor despair ... And where a theorem may present some problem, he may always look down to the numerical examples for help.

PERSPECTIVE PAINTERS

Perspective, a plate from a German book of 1610

Connections between mathematics and the visual arts have been apparent since earliest times – from examples of geometrical cave art and the mathematical decorations on vases and basketry to the geometrical patterns used by the Romans for their mosaics and the Arabs for their tilings.

A notable innovation of early Italian Renaissance painting was that artists became interested in depicting three-dimensional objects in a realistic way, thereby adding visual depth to their works. This soon led to the formal study of geometrical perspective.

The first person to experiment with perspective was the artisan–engineer Filipo Brunelleschi (1377–1446), the designer of the self-supporting octagonal cupola of the cathedral in Florence. Brunelleschi's ideas were developed by his friend Leon Battista Alberti (1404–1472), who presented mathematical rules for correct perspective painting and stated in his *Della Pittura* (On Painting) of 1436 that

> *The first duty of a painter is to know geometry.*

PIERO DELLA FRANCESCA

Another 15th-century Italian artist, Piero della Francesca (c.1415–1492), found a perspective grid useful for his explorations into solid geometry and investigated mathematical perspective more thoroughly than his predecessors. He wrote the treatises *De Prospectiva Pingendi* (On Perspective for Painting) and *Libellus de Quinque Corporibus Regularibus* (Book on the Five Regular Solids).

Della Francesca's *Madonna and Child with Saints,* with lines of perspective

Dürer's guide to drawing in perspective

One of della Francesca's most famous paintings, *Madonna and Child with Saints* (1472), illustrates his mastery of perspective: for example, the parallel lines in the ceiling appear to converge and the squares are not painted as squares. The focal point of the picture is the head of the Madonna.

ALBRECHT DÜRER

The use of perspective was also being investigated elsewhere. As a young man, the celebrated German artist and engraver Albrecht Dürer (1471–1528) had travelled to Italy to learn the secrets of perspective, before returning to his homeland to introduce it there. To this end, he produced a number of engravings and woodcuts that showed how to realize perspective in practice, such as the woodcut above.

In one of his most famous engravings, *St. Jerome in his Cell* (1513–14), we see an excellent example of geometrical perspective, where he took an approach that differed from that used in Italy. Here, the vanishing point is close to the right edge of the picture, which consequently seems uncramped and cosy. Each object is carefully placed with perspective in mind, and appears at right angles to the picture, parallel to it, or at a 45° angle; for example, one of the slippers on the left has been placed parallel to the wall while the other is perpendicular to it.

A mathematician with much original work to his credit, Dürer invented machines for perspective drawing, discovered new geometrical curves and polyhedra, and wrote for the practitioner on the use of mathematics in construction and design. His books were studied widely and inspired a great deal of later mathematical activity.

Dürer's *St Jerome in his Cell*

PACIOLI AND DA VINCI

Johann Gutenberg's invention of the printing press (around 1440) revolutionized mathematics, enabling classical mathematical works to be accessible for the first time. Previously, scholarly works, such as the classical texts of Euclid and Archimedes, had been available only to scholars in manuscript form, but the printed versions made these works much more widely available, just as the internet does today.

At first the new books were printed in Latin or Greek for the scholar, and many such editions appeared. The earliest printed version of Euclid's *Elements* was published in Venice in 1482, and ten years later there appeared an attractive 1492 edition of Ptolemy's *Almagest.*

Although these editions were initially in Latin or Greek, vernacular works (those in the language of its readers) began to appear. These included introductory texts in arithmetic, algebra and geometry, as well as practical manuals designed to prepare young men for a commercial career.

The invention of printing also led to a gradual standardization of mathematical notation. In particular, the symbols + and – made their first appearance in 1489 in a German arithmetic text, *Behende und hubsche Rechenung auff allen Kauffmanschafft* (Nimble and Neat Calculation in all Trades), by Johannes Widmann. Surprisingly, the symbols × and ÷ did not come into general use until the 17th century.

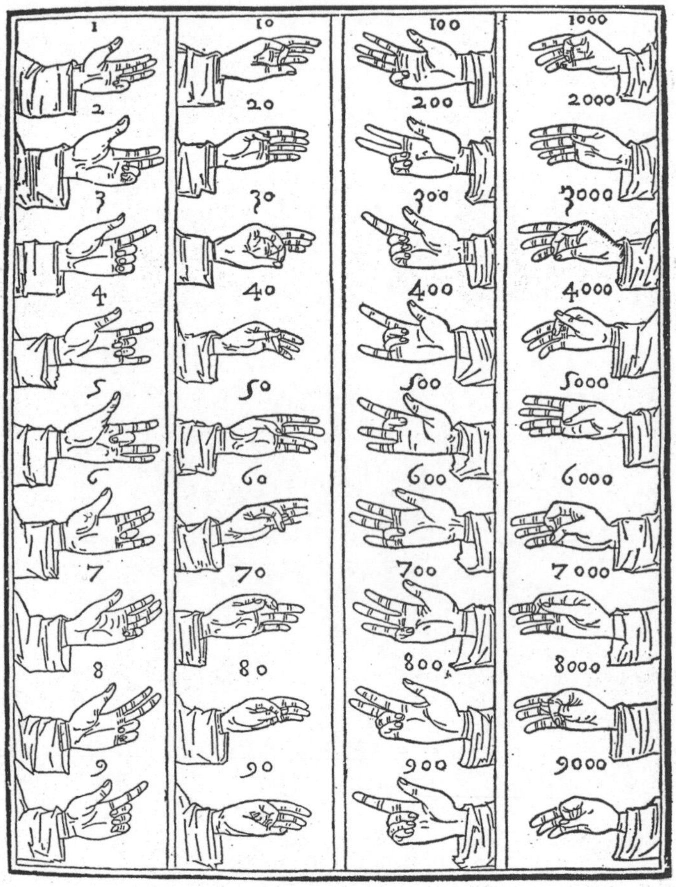

Finger counting, from Pacioli's *Summa*

LUCA PACIOLI

An important and influential vernacular text, first published in 1494, was the *Summa de arithmetica, geometrica, proportioni et proportionalita* (Summary of Arithmetic, Geometry, Proportion and Proportionality), by Luca Pacioli (1447–1517), a mathematics teacher and Franciscan friar. This is a 600-page compilation of the mathematics known at the time, written in Italian for his students. It is now remembered mainly for including the first published account of double-entry bookkeeping – with the result that Pacioli is sometimes called 'the Father of Accounting'.

Luca Pacioli was a good friend of the painter Piero della Francesca and appears (second from the right) in Piero's *Madonna and Child with Saints.* Another well-known painting featuring Pacioli shows him demonstrating a proposition of Euclid, with a copy of the *Elements* and a pair of compasses on his desk and a polyhedron (a rhombicuboctahedron) suspended from the ceiling.

Luca Pacioli

Another influential work of Pacioli was his *De Divina Proportione* (On Divine Proportion) (1509). The woodcuts of polyhedra for this book were prepared by his friend and student Leonardo da Vinci.

LEONARDO DA VINCI

Leonardo da Vinci (1452–1519) explored perspective as thoroughly as any other Renaissance painter and his notebooks contain much of mathematical interest.

While in service as painter and engineer to the Duke of Milan, advising on architecture, hydraulics and military affairs, Leonardo became fascinated with geometry. He studied Euclid's *Elements* and Pacioli's *Summa* and also read Alberti's writings on architecture and della Francesca's treatise on perspective. It is said that he became so involved with his geometrical pursuits while collaborating with Pacioli that he neglected his paintings.

His other mathematical activities included writing a book on elementary mechanics, and investigating various approaches to squaring the circle. With his background in engineering, his methods were often mechanical in nature, rather than theoretical.

Leonardo made great use of the golden number when planning the proportions of his paintings. Indeed, in his *Trattato della Pittura* (Treatise on Painting) he warns

> *Let no one who is not a mathematician read my work.*

Leonardo produced drawings of polyhedra to illustrate Pacioli's book, *De Divina Proportione*

RECORDE

In England the earliest published books with any mathematical content were in Latin. These included Cuthbert Tunstall's *De Arte Supputandi* (On the Art of Reckoning) of 1522, the first arithmetic text published in England, and the best of its time. But gradually, vernacular works in English began to appear.

The first arithmetic book to be published in English may have been a work from St Albans in 1537 entitled *An Introduction for to Lerne to Reken with the Pen and with the Counters, after the Trewe Cast of Arismetyke or Awgrym in Hole Numbers, and also in Broken* (the word *awgrym* means mathematics, and *broken* numbers are fractions). But the most important early writer of mathematical textbooks in English was Robert Recorde (1510–1558).

The Caſtle of Knowledge.

The Sphere of Deſtinye.

The wheele of Fortune.

Sphæra Fati

Sphæra Fortunæ.

QVI MODO SCANDIT, CORRVET STATIM.

whoſe gouernour is Knowledge.

whoſe ruler is Ignoraunce.

To KNOWLEDG is this Trophy ſet,
All learninges friendes will it ſupport.
So ſhall their name great honour get,
And gaine great fame with good report

Though ſpitefull Fortune turned her wheele
To ſtaye the Sphere of Vranye,
Yet dooth this Sphere reſiſt that wheele,
And fleeyth all fortunes villanye.
Though earthe do honour Fortunes balle,
And bytells blynde hyr wheele aduaunce,
The heauens to fortune are not thralle,
Theſe Spheres ſurmount al fortunes chance.

Recorde's *The Castle of Knowledge*

ROBERT RECORDE

Recorde had an eventful life. Graduating from Oxford University in 1531, he was elected a Fellow of All Souls, before going to Cambridge to study mathematics and medicine. Later he became General Surveyor of the Mines and Monies in Ireland, until the project was closed down. He then apparently went to London, acting as physician to Edward VI and Queen Mary. When his rival, the Earl of Pembroke, led troops to dispel a rebellion against the Queen, Recorde tried to charge him with misconduct and was sued for libel. Unable or unwilling to pay his fine, he was thrown into a London jail for debt and died there.

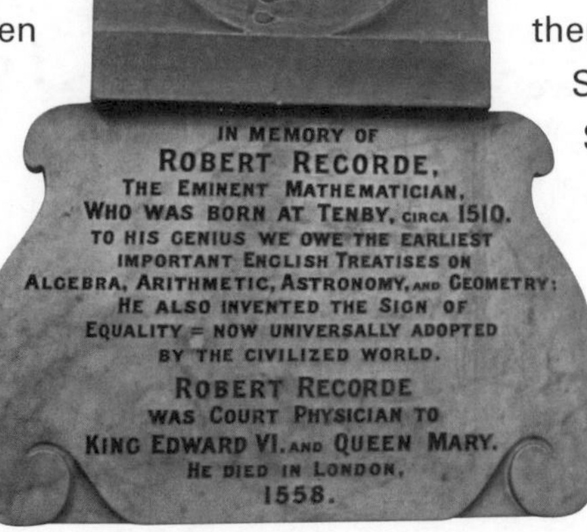

A memorial to Robert Recorde in St Mary's Church, Tenby, South Wales

On the educational side, Recorde was a highly respected communicator. His books, all written in English, were designed to teach mathematics and its applications to the general reader and ran to many editions. Most of them were written in the form of a Socratic dialogue between a Scholar and his Master.

THE GROUND OF ARTES (1543)

The first of Recorde's books, *The Ground of Artes,* was an arithmetic text teaching 'the Worke and Practise, of Arithmeticke, both in whole numbers and fractions' and explaining the various rules so simply that 'euerie child can do it'. In addition to leading the Scholar through the techniques that he needed to learn, the Master explained the importance of arithmetic in everyday life, discussed its uses in commerce and war, and justified its appearance in such 'other Sciences, as Musick, Physick, Law, Grammer and such like'. However, the Master's advice was occasionally less than encouraging:

> Scholar. *Syr, what is the chiefe use of Multiplication?*
>
> Mayster. *The use of it is greater than you can yet understand.*

In this section on multiplication the Master explained how to carry out multiplication sums. To multiply 8 by 7, for example, he wrote these numbers on the left, and opposite he subtracted each from 10, giving 2 and 3.
Now 8 − 3 (or 7 − 2) = 5, and 3 × 2 = 6,
so the answer is 56.

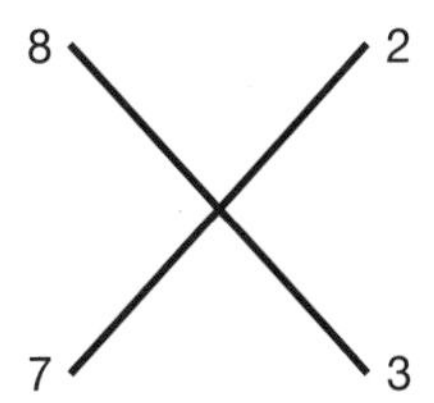

The cross eventually shrank in size and became the multiplication sign that we use today.

The Pathway to Knowledge (1551)

The appearance of printed books led to standardization in terminology. In his geometry text, *The Pathway to Knowledge,* Recorde introduced the term *straight line,* which is still used. He also proposed several attractive terms that never caught on, such as *prickes* for points, *sharp* and *blunt corners* for acute and obtuse angles, a *touch line* for a tangent, *threelike* for an equilateral triangle, and a *likejamme* for a parallelogram.

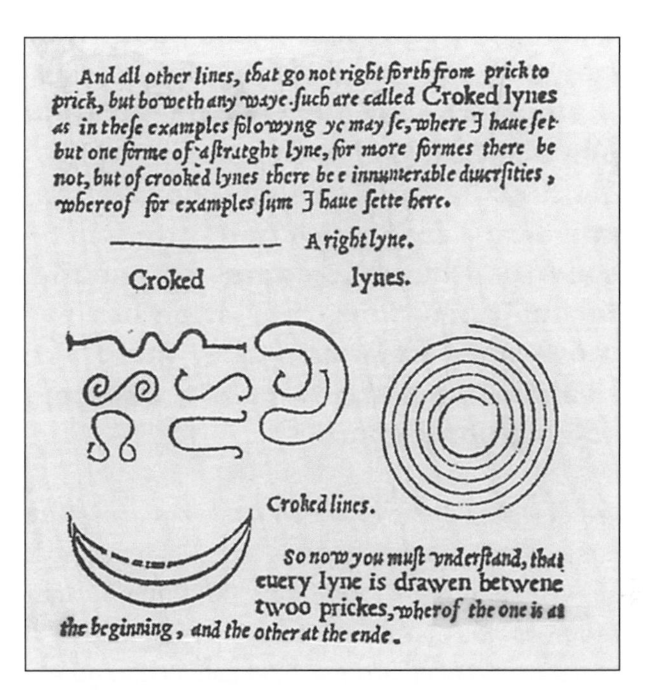

And all other lines, that go not right forth from prick to prick, but boweth any waye, such are called Croked lynes as in these examples folowyng ye may se, where I haue set but one forme of a straight lyne, for more formes there be not, but of crooked lynes there be innumerable diuersities, whereof for examples sum I haue sette here.

A right lyne.

Croked lynes.

Croked lines.

So now you must vnderstand, that euery lyne is drawen betwene twoo prickes, wherof the one is at the beginning, and the other at the ende.

An extract from *The Pathway to Knowledge*

The Whetstone of Witte (1557)

Recorde's most celebrated use of new notation made its first appearance in *The Whetstone of Witte,* his textbook on algebra. In this work he explains that,

> *to avoide the tediouse repetition of these woordes: is equalle to: I will sette as I doe often in woorde use, a paire of paralleles, o: Gemowe lines of one lengthe, thus: ═══, bicause noe.2. thynges, can be moare equalle.*

This was the first appearance of our equals sign; it was much longer than our current version.

1. 14.ze. + .15.ρ ═══ 71.ρ.
2. 20.ze. — .18.ρ ═══ .102.ρ.
3. 26.ʒ. + 10ze ═══ 9.ʒ. — 10ze + 213.ρ.

Recorde's other books included a book on medicine called *The Urinal of Physick* (1548) and an astronomy text entitled *The Castle of Knowledge* (1556).

CARDANO AND TARTAGLIA

The attempt to solve cubic equations is one of the most celebrated stories in the history of mathematics. It took place in Bologna in the early 16th century, during a period when Italian university academics had little job security. Having to compete annually for their positions, they often had to prove their superiority over their rivals by resorting to public problem-solving contests.

We have seen how Omar Khayyam classified cubic equations and solved one by intersecting a semicircle with a parabola. But little further progress was made on solving cubic equations in general, and even around 1500 Pacioli and others were pessimistic as to whether this could be done.

However, in the 1520s, Scipione del Ferro, a mathematics lecturer at the University of Bologna, found a general method for solving cubic equations of the form

Niccolò Tartaglia

A cube and things equal to numbers

(which we would write as $x^3 + cx = d$), and revealed it to his pupil Antonio Fior.

Another who investigated cubic equations around this time was Niccolò of Brescia (1499/1500–1557), known as Tartaglia ('the stammerer') from a bad stammer that he developed after being slashed by a sabre across the face as a boy. In particular, Tartaglia found a method for solving equations of the form

A cube and squares equal to numbers

(which we would write as $x^3 + bx^2 = d$).

FIOR CHALLENGES TARTAGLIA

After del Ferro's death in 1526, Fior felt free to exploit his secret, and challenged Tartaglia to a cubic-solving contest. Fior presented him with thirty cubic equations of the first form, giving him a month to solve them. Tartaglia in turn presented Fior with thirty cubic equations of the second form.

Here are two of Fior's problems, with their modern equivalents:

Find me a number such that when its cube root is added to it, the result is 6. [$x^3 + x = 6$]

A man sells a sapphire for 500 ducats, making a profit of the cube root of his capital. How much is this profit? [$x^3 + x = 500$]

Fior lost the contest. He was not a good enough mathematician to solve Tartaglia's type of problem, while Tartaglia, during a sleepless night ten days before the contest, managed to discover a method for solving all of Fior's problems.

GEROLAMO CARDANO

Meanwhile in Milan, Gerolamo Cardano (1501–1576) was writing extensively about a range of topics, from physics and medicine to algebra and probability (especially its applications to gambling). On hearing about the contest, Cardano determined to prise Tartaglia's method out of him.

TARTAGLIA'S METHOD FOR SOLVING $x^3 + cx = d$

In order to keep it secret, Tartaglia memorized his method in the form of a verse. This appears in italics below, together with the general method and a solution in a particular case — the equation $x^3 + 18x = 19$, where $c = 18$ and $d = 19$.

The method involves finding two numbers u and v satisfying $u - v = d$ and $uv = (c/3)^3$ and then writing $x = \sqrt[3]{u} - \sqrt[3]{v}$.

When the cube and the thing together
Are equal to some discrete number, [$x^3 + cx = d$: $x^3 + 18x = 19$]
Find two other numbers differing in this one. [$u - v = d$: $u - v = 19$]
Then you will keep this as a habit
That their product shall always be equal
Exactly to the cube of a third of the things. [$uv = (c/3)^3$: $uv = 6^3 = 216$]
The remainder then as a general rule [Find u, v: $u = 27$, $v = 8$]
Of their cube roots subtracted [Find $\sqrt[3]{u}$, $\sqrt[3]{v}$: $\sqrt[3]{u} = 3$, $\sqrt[3]{v} = 2$]
Will be equal to this principal thing. [$x = \sqrt[3]{u} - \sqrt[3]{v}$: $x = 3 - 2 = 1$, so $x = 1$]

This he did one evening in 1539, after promising Tartaglia an introduction to the Spanish Governor of the city. Tartaglia hoped that the Governor would fund his researches, and in turn extracted from Cardano the following solemn oath not to reveal his method of solution:

I swear to you, by God's holy Gospels, and as a true man of honour, not only never to publish your discoveries, if you teach me them, but I also promise you, and I pledge my faith as a true Christian, to note them down in code, so that after my death, no-one will be able to understand them.

However, in 1542 Cardano learned that the original discovery of Tartaglia's method had been due to del Ferro, and he felt free to break his oath. Meanwhile, his brilliant colleague, Ludovico Ferrari, had found a similar general method for solving quartic equations (involving terms in x^4).

In 1545, Cardano published *Ars Magna* (The Great Art), containing the methods for solving cubics and quartics and giving credit to Tartaglia. The *Ars Magna* became one of the most important algebra books of all time, but Tartaglia was outraged by Cardano's behaviour and spent the rest of his life writing him vitriolic letters.

Thus, after a struggle lasting many centuries, cubic equations had at last been solved, together with quartic equations. The next question (*Can one solve equations involving x^5, x^6,...?*) remained open until the 19th century.

Gerolamo Cardano

BOMBELLI

For most purposes our ordinary number system is all we need. In this system we can take the square root of numbers such as 3, $\sqrt{2}$ and π, but not the square root of the negative number -1: this is because both positive and negative numbers have positive squares, so what number can one square to give -1? Cardano and Bombelli came across this problem while trying to solve quadratic and cubic equations, and they found it useful to calculate with the mysterious object $\sqrt{-1}$, even though they did not understand what it was.

One numerical problem that Cardano tried to solve was:

> *Divide 10 into two parts whose product is 40.*

On taking the parts to be x and $10 - x$, he obtained the quadratic equation

$$x(10 - x) = 40.$$

Solving this, he obtained the solutions $5 + \sqrt{-15}$ and $5 - \sqrt{-15}$. He could see no meaning to these, but observed

> *Nevertheless we will operate, putting aside the mental tortures involved,*

and found that everything worked out correctly:

- the sum is $(5 + \sqrt{-15}) + (5 - \sqrt{-15}) = 10$;
- the product is $(5 + \sqrt{-15}) \times (5 - \sqrt{-15}) = 5^2 - (\sqrt{-15})^2 = 25 - (-15) = 40$.

In view of these 'mental tortures', Cardano was led to complain that:

> *So progresses arithmetic subtlety, the end of which is as refined as it is useless.*

RAFAEL BOMBELLI

The situation was greatly clarified by Rafael Bombelli (c.1526–1572), who was born in Bologna and later worked as an engineer, draining swampy marshes and reclaiming them for the Catholic Church.

Growing up in Bologna, Bombelli was aware of the dispute between Cardano and Tartaglia and became interested in cubic equations and how to solve them. In particular, he considered the equation $x^3 = 15x + 4$, which has three real solutions:

$$x = 4, \quad -2 + \sqrt{3} \quad \text{and} \quad -2 - \sqrt{3},$$

with not an imaginary number in sight. However, on applying Tartaglia's method to solve this cubic equation, Bombelli was surprised to obtain the solution

$$x = \sqrt[3]{(2 + \sqrt{-121})} + \sqrt[3]{(2 - \sqrt{-121})},$$

which involves complex numbers.

L'ALGEBRA
OPERA
Di RAFAEL BOMBELLI da Bologna
Diuiſa in tre Libri.
Con la quale ciaſcuno da ſe potrà venire in perfetta cognitione della teorica dell'Arimetica.
Con vna Tauola copioſa delle materie, che in eſſa ſi contengono.
Poſta hora in luce à beneficio delli ſtudioſi di detta profeſſione.

IN BOLOGNA,
Per Giouanni Roſſi. MDLXXIX.
Con licenza de' Superiori.

Title page of Bombelli's *Algebra*

COMPLEX NUMBERS

Suppose that we try to calculate with the symbol $\sqrt{-1}$.

We find that addition is easy:

$$(2 + 3\sqrt{-1}) + (4 + 5\sqrt{-1}) = 6 + 8\sqrt{-1},$$

and so is multiplication (replacing $\sqrt{-1} \times \sqrt{-1}$ whenever it appears by -1):

$$(2 + 3\sqrt{-1}) \times (4 + 5\sqrt{-1}) = (2 \times 4) + (3\sqrt{-1} \times 4) + (2 \times 5\sqrt{-1}) + (15 \times \sqrt{-1} \times \sqrt{-1})$$
$$= (8 - 15) + (12 + 10)\sqrt{-1} = -7 + 22\sqrt{-1}.$$

We can carry out all the standard operations of arithmetic on these new objects. We call the object $a + b\sqrt{-1}$ a *complex number*: the number a is its *real part*, and the number b is its *imaginary part*. Nowadays, we usually use the letter i to mean $\sqrt{-1}$, so that $i^2 = -1$.

In 1799 complex numbers were given a geometrical form by the Danish navigator Caspar Wessel. In this representation, called the *complex plane*, two axes are drawn at right angles (the *real axis* and the *imaginary axis*) and the complex number $a + b\sqrt{-1}$ is represented by the point at distance a in the direction of the real axis and height b in the direction of the imaginary axis.

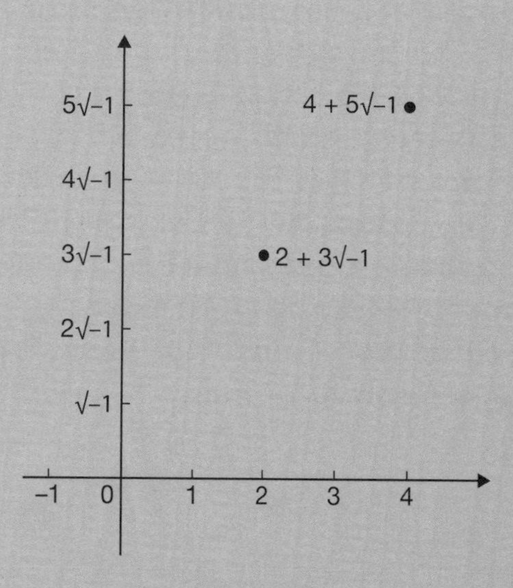

Representing complex numbers

To explain the connection between his solution and the solution $x = 4$, Bombelli looked for real numbers a and b such that

$$(a + b\sqrt{-1})^3 = 2 + \sqrt{-121}$$
$$\text{and } (a - b\sqrt{-1})^3 = 2 - \sqrt{-121},$$

so that he could take the two cube roots. After some experimentation, he found that $a = 2$ and $b = 1$:

$$(2 + \sqrt{-1})^3 = 2 + \sqrt{-121}$$
$$\text{and } (2 - \sqrt{-1})^3 = 2 - \sqrt{-121},$$

and so $x = (2 + \sqrt{-1}) + (2 - \sqrt{-1}) = 4$, as expected.

BOMBELLI'S *ALGEBRA*

Throughout his life Bombelli studied the algebraic writings of his predecessors, such as al-Khwarizmi, Fibonacci and Pacioli. He also embarked on a major study of the works of Diophantus, having been shown a copy of the *Arithmetica* in a library in Rome.

Believing that no-one had really explained clearly the nature of algebraic problems, and of how to solve cubic equations in particular, Bombelli embarked on a major project in which he proposed to present all that was known about the subject in an accessible form. Although five volumes of his *Algebra* were planned, only three were completed before his death. Incomplete manuscripts of the final two volumes (the 'geometrical part') were discovered in a Bologna library in 1923.

In his *Algebra* Bombelli described how he had struggled hard to understand complex numbers. He was the first to show how to add and subtract them, and he gave rules for multiplying them. Using these rules, Bombelli showed how to obtain real solutions of cubic equations, even when Tartaglia's method yields square roots of negative numbers.

MERCATOR

The Flemish mapmaker and cartographer Gerardus Mercator (1512–1594) is mainly remembered for the *Mercator projection,* which proved to be extremely useful for navigators. This was a projection of the spherical earth on to a flat sheet of paper so that the lines of latitude and longitude, as well as the paths of constant compass bearings, were represented by straight lines. Mercator also coined the word 'atlas' for a collection of maps.

A major concern in the 16th century, an active period for voyages of trade, discovery and exploration, was to develop mathematical methods and maps to aid navigation.

The basic problem was that if you were on a ship in the middle of the ocean, how could you tell where you were, and in which direction you should sail to get to your destination? You could find your latitude by using astronomical instruments to locate the sun and stars. It was more problematic, however, to find your longitude, and a satisfactory method was not available until the end of the 18th century.

Mercator (left) and Jodocus Hondius, who published Mercator's work, on the title page of an edition of the *Mercator–Hondius Atlas,* surrounded by the tools of the cartographer.

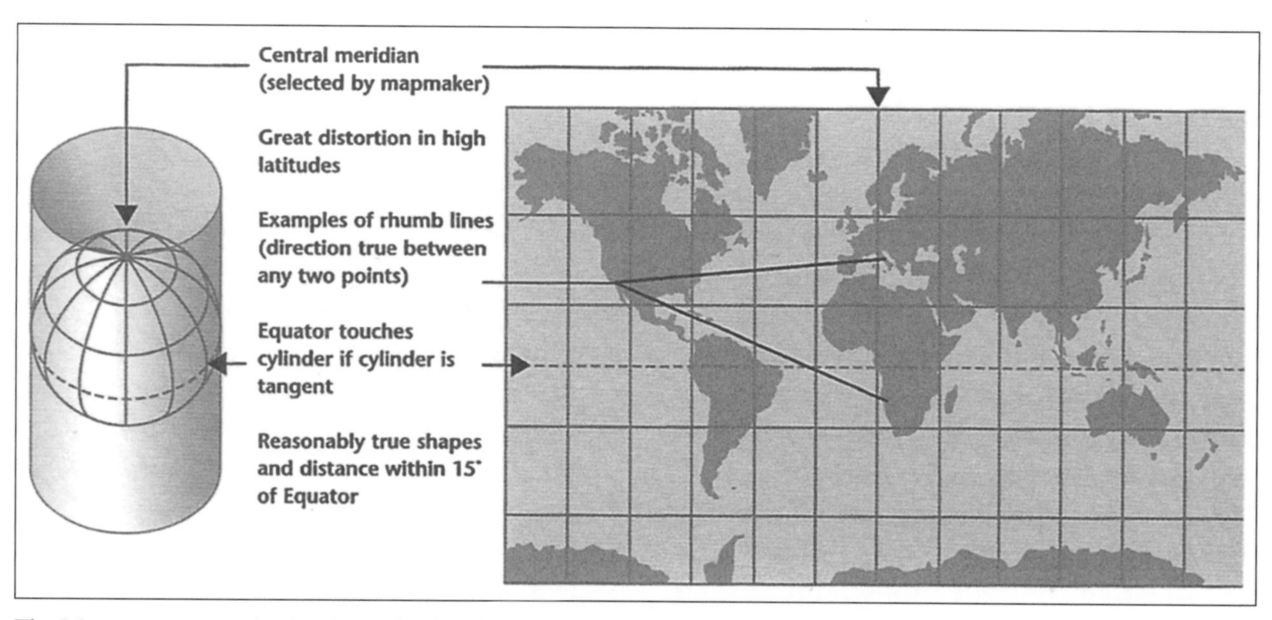

The Mercator map projection is a cylindrical one that distorts the size and shape of areas far from the equator.

Using a magnetic compass, mariners could steer a line of constant compass bearing (a *rhumb line*); such a path crosses all lines of longitude at the same angle. However, as the 16th-century Portuguese mathematician and cosmographer Pedro Nunes discovered, a rhumb line spirals towards the pole.

MERCATOR'S PROJECTION

The advantages of Mercator's projection were that it represented lines of latitude and longitude as straight lines meeting at right angles, and that it also represented rhumb lines as straight lines on the map. If a navigator knew the latitudes and longitudes of the ship's current position and the destination, then the line joining the two places could be found on the map. This enabled the appropriate constant compass bearing to be determined, but would not give the shortest distance to the destination.

Mercator's projection marked a step forward for navigation

Mercator obtained his projection by projecting the sphere on to a cylinder which was then unrolled and stretched vertically so that the rhumb lines became straight; the amount of stretching varies with different latitudes, and increases the further north one goes. This has the consequence of exaggerating areas that are far from the equator: for example, Alaska appears as large as Brazil while Brazil is actually five times bigger, and Finland has a greater north–south extent than India which is incorrect.

Mercator did not present the mathematical basis for his projection. This was first given by Edward Wright in his influential *Certaine Errors in Navigation* of 1599. Wright also gave accurate mathematical tables for its construction. But it was Thomas Harriot (1560–1621) who eventually discovered the fundamental mathematical formula underlying Mercator's projection.

COPERNICUS AND GALILEO

Nicolaus Copernicus (1473–1543), the 'Father of Modern Astronomy', was born in Torun in Poland, and studied in Cracow, Bologna and Ferrara. He transformed his subject by replacing Ptolemy's earth-centred system of planetary motion by a heliocentric system with the sun at the centre and the earth as just one of several planets travelling in circular orbits around it. In 1632, Galileo Galilei (1564–1642), the 'Father of Modern Physical Science', described the advantages of the Copernican system over Ptolemy's, bringing him into trouble with the Inquisition. In 1638 he wrote a book on mechanics that set the scene for the work of Isaac Newton and others.

NICOLAUS COPERNICUS

Although the heliocentric idea had been suggested earlier by Aristarchus and others, Copernicus was the first to work out the underlying theory and its consequences in mathematical detail.

His book *De Revolutionibus Orbium Coelestium* (On the Revolutions of the Heavenly Spheres) was published in 1543 and a copy was reportedly presented to him as he lay on his deathbed. In this celebrated work, he showed that the six planets then known split into two groups: Mercury and Venus (with orbits inside that of Earth), and Mars, Jupiter and Saturn (with orbits outside it). He listed these planets in increasing order of distance from the sun, and thereby illuminated phenomena that the Ptolemaic system had failed to explain, such as why Mercury and Venus are visible only at dawn and dusk while the other planets are visible throughout the night.

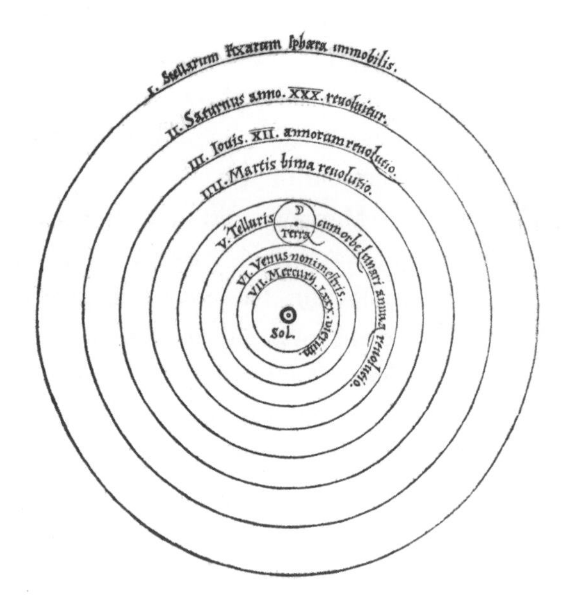

Copernicus's heliocentric system, from *De Revolutionibus Orbium Coelestium*

GALILEO GALILEI

The Copernican solar system aroused much controversy and brought its supporters into direct conflict with the Church who considered

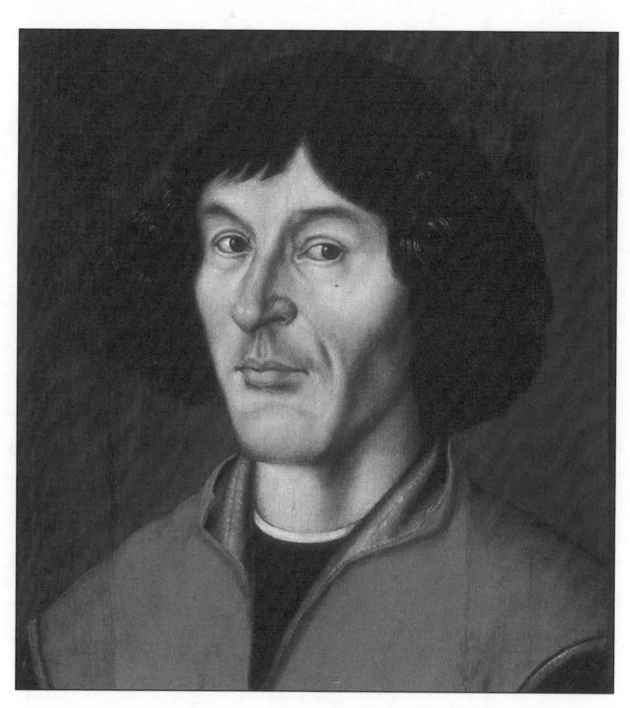

Nicolaus Copernicus

Frontispiece of the *Two Chief World Systems,* depicting Aristotle, Ptolemy and Copernicus in animated conversation

the earth to lie at the centre of Creation, so that Copernicus's ideas were at variance with the Holy Scriptures. It was Galileo who gave the most compelling exposition of the Copernican theory in his *Dialogo sopra I Due Massimi Sistemi del Mondo* (Dialogue Concerning the Two Chief World Systems) (1632). This led to his trial by the Inquisition when he was forced to recant his Copernican views. He was not pardoned by the Church until 1995.

Galileo was born in Pisa and taught mathematics in Padua from 1592 to 1610, after which he became mathematician and philosopher to the Grand Duke at Florence. He was one of the first astronomers to make extensive use of the telescope, discovering sunspots and the moons of Jupiter and drawing the moon's surface. This instrument also enabled him to obtain observational evidence in favour of the Copernican system. In the Ptolemaic system the same amount of Venus would always be visible, whereas in the Copernican system it should show phases (different amounts visible at different times). He turned his telescope on Venus and over the following few nights observed its phases.

Galileo's *Two Chief World Systems* was written in Italian and takes the form of a discussion over four days between two philosophers and a layman. They are Salviati who argues for the Copernican position and presents Galileo's views, Sagredo, the layman, who is seeking the truth and Simplicio, a follower of Aristotle and Ptolemy, who presents the traditional views and arguments.

GALILEO'S MECHANICS

In his mechanics book of 1638, Galileo discussed the laws of uniform and accelerated motion and explained why the path of a projectile must be a parabola.

In this book he gathered together a lifetime of study, presenting a theory of how position, velocity and acceleration vary with time and supporting it with mathematical deductions. It was here that he laid the mathematical foundations underpinning his belief that the earth really moves. Its mathematical form laid a foundation for further advance by others, and particularly by Isaac Newton who was born in the year that Galileo died.

KEPLER

Johannes Kepler (1571–1630) was born in Swabia in south-west Germany. A gifted and well-read mathematician in the neoplatonist tradition, he based his work on harmony and design. A powerful early expression of this was his model of the solar system, and subsequently the three planetary laws that bear his name. Kepler also investigated polyhedra and contributed to what became the integral calculus.

KEPLER'S CALCULUS

Kepler was interested in what was later known as the *integral calculus,* in which areas and volumes of geometrical shapes are calculated. To obtain such areas and volumes he used what came to be called 'the method of infinitesimals'. By dividing a volume into very thin discs, for example, he determined the volumes of over ninety solids obtained by rotating conics and other curves around an axis.

The Danish astronomer Tycho Brahe (1546–1601) was the greatest observer of the heavens before the invention of the telescope, working for many years at his observatory of Uraniborg on the Danish island of Hven, before moving to Prague. Kepler became his assistant in Prague and in 1601 was appointed Imperial Mathematician to succeed Brahe after Brahe's untimely death. Kepler spent the next eleven years in Prague and produced some of his most important work there.

Johannes Kepler

KEPLER'S LAWS OF PLANETARY MOTION

What was needed to support the Copernican theory was a method of calculating heavenly events with at least as much accuracy as the Ptolemaic theory had done with its apparatus of epicycles. Using Brahe's extensive observational records, Kepler was eventually led, in his *Astronomia Nova* (The New Astronomy) (1609) and *Harmonices Mundi* (Harmony of the World) (1619), to the three laws that enabled him to make these calculations:

1: *The planets move in elliptical orbits with the sun at one focus.*

2: *The line from the sun to a planet sweeps out equal areas in equal times.*

3: *The square of a planet's period is proportional to the cube of its orbit's mean radius.*

The following diagram illustrates Kepler's first and second laws. It shows a planet in elliptical orbit around the sun which is at a focus of the ellipse, and depicts the path travelled by the planet during three equal periods in its orbit. Kepler's second law tells us that these shaded areas are equal.

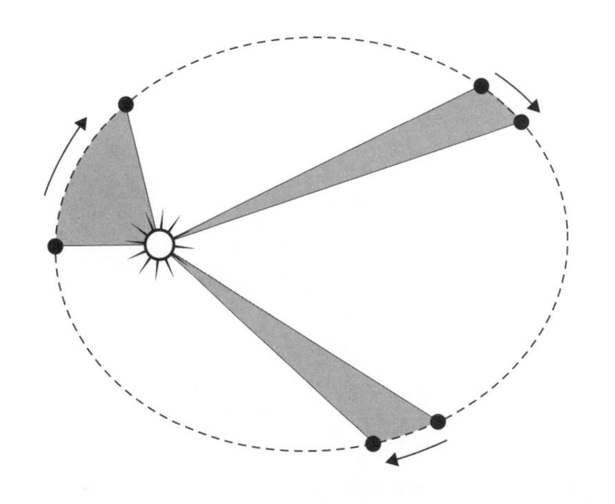

Kepler's laws were based on observed results; it was to be Newton, some eighty years later in his *Principia Mathematica,* who explained why they are true.

MATHEMATICAL PURSUITS

In his *Mysterium Cosmographicum* (The Cosmographic Mystery) of 1596, Kepler proposed a model of the solar system in which the five Platonic solids were set inside each other (with the octahedron innermost, followed by the icosahedron, dodecahedron, tetrahedron and cube) and then interspersed with six spheres that carried the orbits of the then-known planets (Mercury, Venus, Earth, Mars, Jupiter, and Saturn).

Kepler was also interested in polyhedra in general, discovering the cuboctahedron and the antiprisms, and his name is associated with the four *Kepler–Poinsot star polyhedra.*

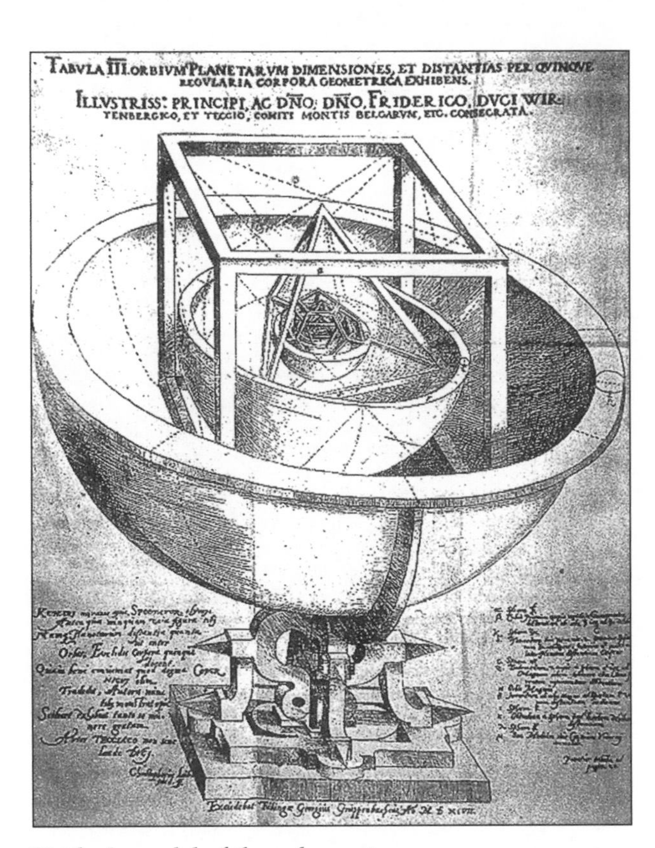

Kepler's model of the solar system

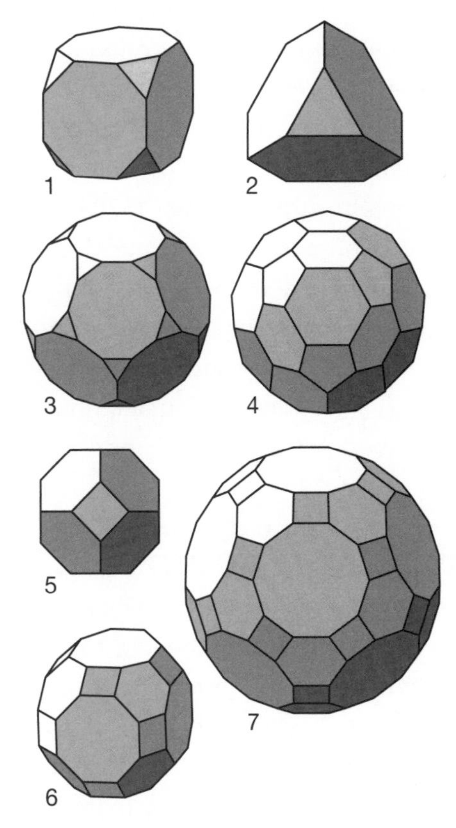

Some of Kepler's drawings of polyhedra

VIÈTE

François Viète (1540–1603), sometimes known as Vieta (the Latin form of his name), was a French mathematician who was born at Fontenay-le-Compte and died in Paris. He trained as a lawyer and served as a counsellor to both Henry III and Henry IV. His work on notation and algebraic modes of thought were fundamental in the development of algebra as an independent branch of mathematics.

Viète is commemorated on this Paris street sign in the 17th arrondissement.

After graduating in law from the University of Poitiers, Viète practised at the bar before serving as a tutor and secretary in the family of Antoinette d'Aubeterre. It was during this time that he was first involved in mathematics, writing mainly on astronomy and trigonometry.

During this period Viète began to write his *Canon Mathematicus seu Ad Triangula* (Mathematical Laws, or On Triangles), which was published throughout the 1570s and which developed and systematized the Greek study of plane and spherical triangles. It also introduced a new notation for decimal fractions and included tables of all six trigonometric ratios for each minute of arc, calculated using inscribed and circumscribed polygons with up to thousands of sides.

VIÈTE'S *ANALYTIC ART*

Viète's reputation is based mainly on the work that he undertook during the late 1580s, when he wrote primarily on algebra and geometry.

His chief legacy was a reversal of the previous situation where geometry was used to justify algebra, and his work inspired many succeeding mathematicians. Algebra was by now obtaining a power that would lead to its replacing geometry as the main language of mathematics. This would allow new insights into existing problems and the development of new areas of mathematical investigation.

Viète's *In Artem Analyticem Isagoge*

François Viète, from Van Schooten's 1646 edition of his collected works.

VAN ROOMEN'S CHALLENGE

In 1594 the Belgian mathematician Adriaan van Roomen challenged the mathematical world to solve the following complicated polynomial equation:

$$x^{45} - 45x^{43} + 945x^{41} - 12300x^{39} + 111150x^{37} - 740459x^{35} + 3764565x^{33} - 14945040x^{31} + 469557800x^{29} - 117679100x^{27} + 236030652x^{25} - 378658800x^{23} + 483841800x^{21} - 488494125x^{19} + 384942375x^{17} - 232676280x^{15} + 105306075x^{13} - 34512074x^{11} + 7811375x^{9} - 1138500x^{7} + 95634x^{5} - 3795x^{3} + 45x = C,$$

where $C = {}^1/_2\sqrt{\{7 - \sqrt{5} - \sqrt{(15 - 6\sqrt{5})}\}}$.

Viète found a solution in just a few minutes, having noticed that the problem is related to that of expressing sin 45*x* in terms of sin *x*.

(Introduction to the Analytic Art) of 1591 was the first of his works on algebra. Here he introduced a new kind of analysis for discovery in mathematics, concerned with solving equations and proportions. In other treatises he solved several equations up to the fourth degree.

VIÈTE'S NOTATION

In his book Viète also pioneered an improvement in notation, using letters for quantities, rather than writing them in words. He denoted all known quantities by consonants (*B*, *D*, etc.) and all unknown quantities by the vowels (*A*, *E*, etc.)

Viète insisted on the importance of dimension, disallowing earlier problems that had involved 'adding lines to areas' (meaning $ax^2 + bx$). He also insisted on preserving homogeneity although he did not restrict himself to three dimensions. The signs + and − appear in his work but not × or ÷. He would write, for example,

A cubus, + A quadrato in B ter,

+ A in B quadratum ter, + B cubo

which we would write in modern notation as

$A^3 + 3A^2B + 3AB^2 + B^3$.

Viète's use of letters was later extended by Descartes.

CIRCLE MEASUREMENT

Viète was interested in the ratio of the circumference of a circle to its diameter, now called π. By calculating with polygons with 393,216 sides, he found π to ten decimal places.

Around the same time, Adriaan van Roomen used polygons with 1,073,741,824 sides to find π to fifteen decimal places, and Ludolph van Ceulen used polygons with about 4×10^{18} sides to extend this to thirty-five decimal places (which he arranged to have carved on his gravestone).

Viète also produced the first exact expression for π, involving many square roots and given below; this arose from his formula

$2/\pi = \cos \pi/2 \times \cos \pi/4 \times \cos \pi/8 \times \ldots$,

where the angles are all measured in radians (with π radians = 180°). His exact expression for π was

$$2 \times \frac{2}{\sqrt{2}} \times \frac{2}{\sqrt{2+\sqrt{2}}} \times \frac{2}{\sqrt{2+\sqrt{2+\sqrt{2}}}} \times \cdots$$

HARRIOT

Thomas Harriot (1560–1621) was arguably England's best mathematical scientist before Isaac Newton, but is now largely unknown because he shared his discoveries only with a small group of friends. His most original work was in navigation and algebra. As scientific and navigational adviser to Walter Ralegh, he contributed important work in map projection.

Harriot left over 8000 manuscript pages of his researches into geometry, algebra, optics, mechanics, astronomy and navigation. He was also the first astronomer to use a telescope to map the moon. Although he was so innovative and original, his work had less impact than it might have had because he did not publish it.

Part of Harriot's *Treatise on Equations*

Recent studies of his algebraic manuscripts have shown what an original thinker Harriot was in his investigation of polynomials — these are expressions like

$$3x^4 - 5x^3 + 2x^2 + 7x - 9,$$

obtained by adding and subtracting numbers and powers of an unknown.

Important for Harriot's success was his notation, which differed from Viète's. He used *ab* to denote *a* multiplied by *b,* and then *aa, aaa* for what we now write as a^2, a^3, etc. He retained Viète's use of vowels for unknown quantities, but wrote them in lower case (*a, e,* etc.). His notation brought many advantages in the way it allowed Harriot to explore the properties of polynomials in terms of their coefficients. Crucially, he saw the possibility of writing polynomials as products of factors of lower degree.

HARRIOT'S NAVIGATION

After graduating from Oxford in 1580, Harriot was employed by Walter Ralegh who was preparing for his colonizing travels to America, to what would eventually be known as Virginia and North Carolina. Harriot advised Ralegh on navigational matters, astronomy and surveying. In fact, the only book by Harriot to appear in his lifetime

Walter Ralegh

A briefe and true report
of the new found land of Virginia:
of the commodities and of the nature and man
ners of the naturall inhabitants. Discouered by
the English Colony there seated by Sir Richard
Greinuile Knight In the yeere 1585. Which Rema
=ined Vnder the gouernement of twelue monethes,
At the speciall charge and direction of the Honou=
rable SIR WALTER RALEIGH Knight lord Warden
of the stanneries Who therein hath beene fauoured
and authorised by her MAIESTIE
:and her letters patents:
This fore booke Is made in English
By Thomas Hariot seruant to the abouenamed
Sir WALTER, a member of the Colony, and there
imployed in discouering
CVM GRATIA ET PRIVILEGIO CÆS. MA^TIS SPECIA^LI

FRANCOFORTI AD MOENVM
TYPIS IOANNIS WECHELI, SVMTIBVS VERO THEODORI
DE BRY ANNO CIↃ D XC.
VENALES REPERIVNTVR IN OFFICINA SIGISMVNDI FEIRABENDII

Title page of Harriot's report on Virginia

was *A Brief and True Report of the New Found Land of Virginia,* arising from his 1585–86 expedition to North America.

HARRIOT'S WORK ON SPIRALS

Arising from his navigational studies, Harriot was led to investigate a type of spiral known as an *equiangular spiral.*

He first approximated it by a polygon that spirals in by a succession of straight lines, each of which cuts a line from a fixed point (P in the diagram) at the same angle α. He then used a 'cut-and-paste' operation to reconstruct this polygon as a triangle.

If we now shrink the line segments, we obtain polygon spirals that approximate ever more closely the original equiangular spiral. Their areas therefore equal that of the triangle, and their lengths are that of the sum of two sides of the triangle, both of which are easily calculated. This was an amazing result, as it was generally believed that the length of such a curve could not be found.

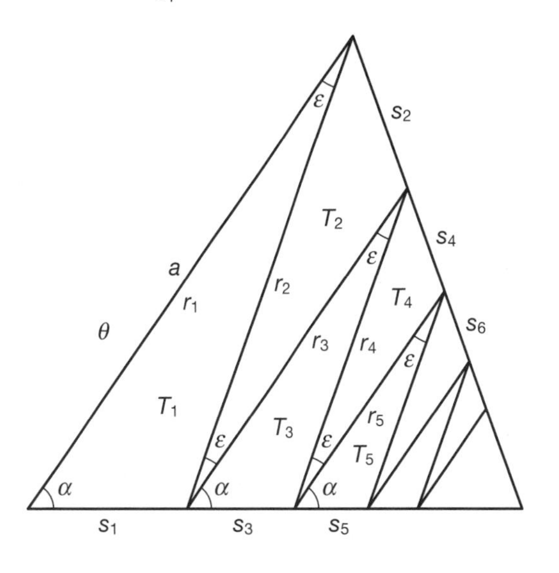

MERSENNE AND KIRCHER

The 13th-century Catalan mystic and poet Ramón Lull believed that all knowledge can be constructed from a small number of basic principles. By taking mathematical combinations of these, he proposed to unify everything known into a single system, and thereby to teach Christian theology so logically that infidels would see its truth and be converted. Lull's influence extended widely and over several centuries, and included such initial converts as the young Leibniz. Two of its strongest adherents in the 17th century were the Minimite friar Marin Mersenne (1588–1648) and the Jesuit priest Athanasius Kircher (1601–1680).

One of Lull's 'circulatory diagrams', showing the relationships among nine of the divine attributes

Believing all reality to be the embodiment of aspects of the divinity, Lull chose eighteen manifestations of the divine attributes of God as the objects to be combined: these included *bonitas* (goodness), *potestas* (power), *sapientia* (wisdom) and veritas (truth).

MARIN MERSENNE

Marin Mersenne, a Minimite friar who lived near Paris in the early 17th century, believed that scientific discoveries should be made available to all. With this aim, he carried out an extensive correspondence with most of the leading European scientists of his day, acting as a 'clearing house' for new scientific results. He also initiated regular meetings in Paris at which mathematicians could meet to discuss their latest findings; these meetings led in 1666 to the founding of the French Academy of Sciences by Louis XIV.

MERSENNE AND MUSIC

Mersenne was also a keen scientist, who carried out practical

Mersenne's book on universal harmony

MERSENNE PRIMES

Mersenne is mainly remembered for his work on prime numbers — notably, those of the form $2^n - 1$ (now called *Mersenne primes*), such as $2^5 - 1 = 31$ and $2^7 - 1 = 127$.
As he realized, *if $2^n - 1$ is prime, then n is prime,*
but the converse statement is false: 11 is prime, but $2^{11} - 1 = 2047 = 23 \times 89$ is not.
Mersenne found nine of these primes, including $2^{127} - 1$, a 39-digit number.

In Book IX of his *Elements*, Euclid proved that if $2^n - 1$ is prime, then $2^{n-1} \times (2^n - 1)$ is a *perfect number* — that is, it equals the sum of its factors (apart from itself); for example, $2^4 \times (2^5 - 1) = 16 \times 31 = 496$, which is a perfect number. Thus, each Mersenne prime gives rise to a perfect number. Later, Euler proved that all *even* perfect numbers must have this form.

At the time of writing, forty-seven Mersenne primes are known, the largest being $2^{43,112,609} - 1$, which has 12,978,189 digits. When seeking new prime numbers, most searchers look for primes of this form.

experiments on the nature of sound. In particular, he investigated how the note produced by a wire varies with its length, thickness, density and tension, and measured the speed of sound.

In his 1636 book on universal harmony, designed for 'mathematicians and theologians', Mersenne discussed the acoustic properties of many musical instruments.

He also presented some of Lull's combinatorial ideas in a musical setting, exhibiting all the 6! = 720 'songs' that can be formed from six notes. (In general, *n factorial*, written *n*!, is the product of 1, 2, 3, ... and *n*.)

He then listed all the factorial numbers up to 64! (a ninety-digit number), and presented extensive tables of permutations and combinations, noting, for example, that the number of ways of selecting twelve objects from thirty-six is 1,251,677,700.

Kircher's *Ars Magna Sciendi*

ATHANASIUS KIRCHER

Another follower of Lull was the Jesuit priest Athanasius Kircher, a polymath who translated Egyptian hieroglyphics, founded one of the earliest museums, designed magic lanterns, and wrote books on topics ranging from Noah's Ark and China to germs and geology.

Kircher's 1669 *Ars Magna Sciendi sive Combinatoria* (The Great Art of Knowledge, or the Combinatorial Art) presents a system of logic derived from Lull. Of its twelve books, Book III, *Methodus Lulliana* is a general description of Lullian principles and is followed by Book IV, *Ars Combinatoria.* This latter book of nearly fifty pages opens with all the permutations of letters in the words *ORA* and *AMEN.* It has a table of factorials up to 50!, followed by a discussion of how to select various combinations of the eighteen Lullian attributes, including a magnificent table of the 324 ways of combining these attributes in pairs.

DESARGUES

Earlier we saw that the subject of perspective had its origins in early Renaissance painting, being developed by such practitioners as Piero della Francesca and Leonardo da Vinci in Italy and Albrecht Dürer in Germany. In the 16th century such concerns became the province of practical tradesmen whose crafts involved perspective (such as architects and military engineers), as well as of mathematicians such as Girard Desargues (1591–1661), who carried out detailed investigations into the geometry of perspective.

Desargues was born and died in Lyon, in southern France. He was a practical geometer who was involved with such pursuits as the design of sundials, accurate stone-cutting, and the applications of perspective to military engineering.

The Infirmary of the Charity Hospital of Paris, an etching by Abraham Bosse, c.1635, illustrating perspective

TWO PROJECTIVE RESULTS

We look at two celebrated theorems of projective geometry. Desargues theorem was first published in 1648 in a manual on perspective by Abraham Bosse.

PASCAL'S 'HEXAGON THEOREM'

Earlier we saw that Pappus's theorem tells us that if we mark six points on two lines and join them up in a particular way, we produce three new points that must always lie on a straight line. Aged 16, the precocious Blaise Pascal discovered that a similar result holds if we start with six points on any conic (ellipse, parabola or hyperbola). The diagram below shows an illustration of Pascal's theorem when the conic is an ellipse.

Choose six points on it:

A, B, C and P, Q, R.

Now draw the lines

AQ and BP — these meet at a point X

AR and CP — these meet at a point Y

BR and CQ — these meet at a point Z

Pascal's theorem states that, however we choose our initial six points,

The resulting points X, Y and Z always lie on a straight line.

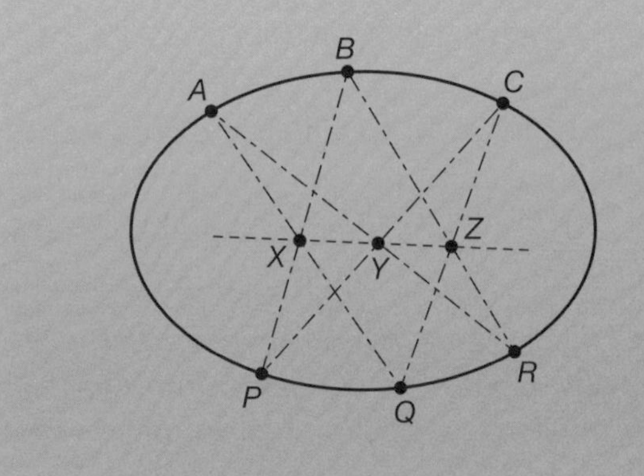

WHAT IS PROJECTIVE GEOMETRY?

Suppose that you draw an equilateral triangle on a piece of glass, and shine light on it from a point source towards a wall. The shadow that appears is still a triangle, but is not an equilateral triangle in general. In fact, by tilting the glass, you can arrange for the shadow to be a triangle of any shape that you wish.

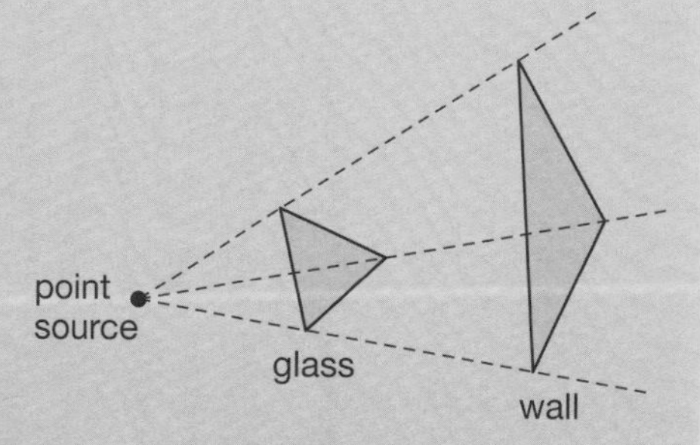

Projective geometry can be thought of as the study of properties that geometrical figures share with their shadows. Here we still have a triangle with straight sides, though no longer an equilateral one in general, so lengths and angles are not preserved.

Desargues discovered an interesting property that always remains unchanged under projections. Take four points *A*, *B*, *C* and *D* on a straight line, measure the lengths *AC*, *AD*, *BC* and *BD*, and calculate the 'ratio of ratios' *AC* / *AD* divided by *BC* / *BD*. Now project these points from a point light source on to another line, giving four new points *A′*, *B′*, *C′* and *D′*, and calculate the ratio *A′C′* / *A′D′* divided by *B′C′* / *B′D′*. Then this 'ratio of ratios' is equal to the earlier one.

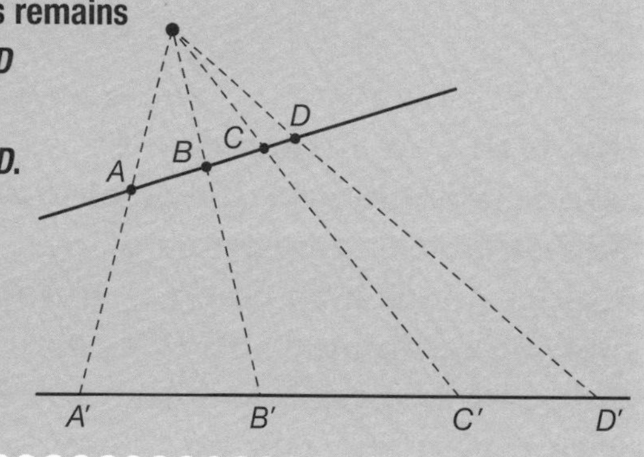

DESARGUES'S THEOREM ON TRIANGLES IN PERSPECTIVE

In the following diagram, the triangles *ABC* and *PQR* are said to be in perspective from the point *O*, since if you look at them from this point, the corresponding pairs of corners *A* and *P*, *B* and *Q*, and *C* and *R*, match up exactly.

Now draw the lines

AB and *PQ* — these meet at a point *X*

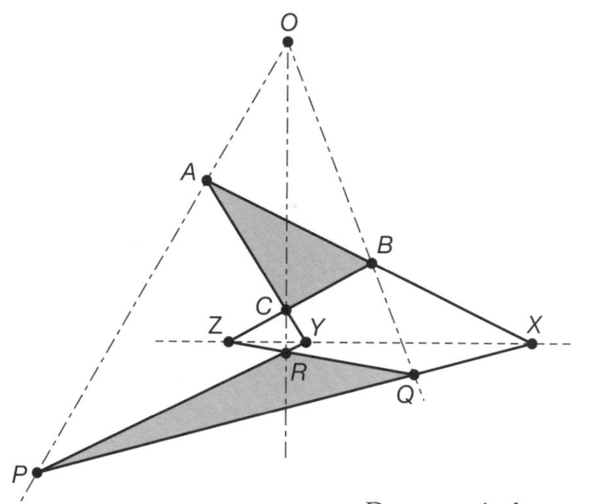

Desargues's theorem

AC and *PR* — these meet at a point *Y*

BC and *QR* — these meet at a point *Z*

Desargues's theorem states that, however we choose our initial two triangles,

The resulting points X, Y and Z always lie on a straight line.

To see why this is true, imagine the above diagram as a picture in three dimensions. The plane containing the triangle *ABC* and the plane containing the triangle *PQR* intersect in the straight line through the points *X*, *Y* and *Z*.

What is striking and unusual about these results is that, unlike the propositions in Euclid's *Elements*, they are entirely about *incidence* — points lying on lines, and lines meeting in points — there is no mention of lengths or angles. Later mathematicians regarded such results as *projective*, since they still hold if we project the figure from a point source of light. Most geometrical theorems do not survive such projections: we lose the Pythagorean theorem, for example, because projecting a right angle does not generally give a right angle.

CHAPTER 3

AWAKENING AND ENLIGHTENMENT

The 17th and 18th centuries witnessed the beginnings of modern mathematics. New areas of the subject came into being – notably, analytic geometry and the calculus – while others, such as number theory, were reborn or took on a new lease of life. Fundamental problems, such as that of determining the orbits of the heavenly bodies, were solved or investigated with novel techniques.

It was the age of Newton in England, Descartes and Pascal in France, and Leibniz in Germany, followed by a succession of Continental 'greats': the Bernoulli brothers, Euler, Lagrange and Laplace.

It was also the age of gatherings – the formation of national scientific societies, such as London's Royal Society and the Academy of Sciences in Paris, and the founding of scholarly institutions such as the St Petersburg Academy and the Academy of Sciences in Berlin.

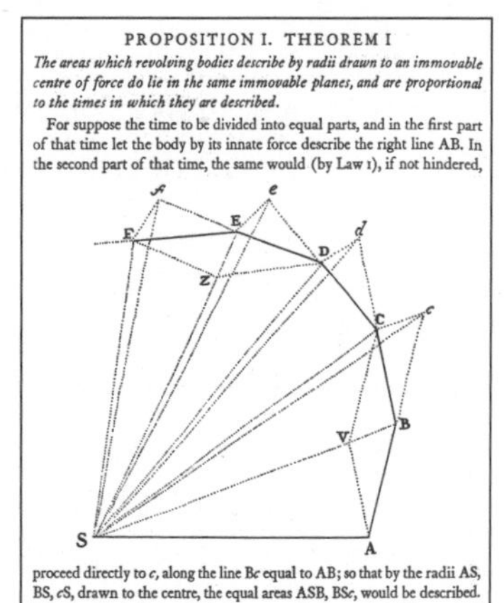

PROPOSITION I. THEOREM I

The areas which revolving bodies describe by radii drawn to an immovable centre of force do lie in the same immovable planes, and are proportional to the times in which they are described.

For suppose the time to be divided into equal parts, and in the first part of that time let the body by its innate force describe the right line AB. In the second part of that time, the same would (by Law 1), if not hindered,

proceed directly to *c*, along the line B*c* equal to AB; so that by the radii AS, BS, *c*S, drawn to the centre, the equal areas ASB, BS*c*, would be described. But when the body is arrived at B, suppose that a centripetal force acts at once with a great impulse, and, turning aside the body from the right line B*c*, compels it afterwards to continue its motion along the right line BC.

The area swept out by a moving body: Newton's use of geometry (right) contrasts with Laplace's analytical approach (far right)

CALCULUS AND DISCOVERY

Initially, the problems that mathematicians solved were geometrical, as were their answers, although the techniques they used (including the calculus) were not necessarily of this kind, being seen as methods of proceeding from a geometrical problem to a geometrical answer. The 18th century then led to a new conception of mathematics, with its most striking characteristic being its algebraic appearance.

The objects of mathematics were now described by formulas with symbols for

> If we project the body m, on the plane of x and y, the differential $(xdy - ydx)/2$, will represent the area which the radius vector, drawn from the origin of the coordinates to the projection of m, describes in the time dt, consequently the sum of the areas, multiplied respectively by the masses of the bodies, is proportional to the element of time, from which it follows that in a finite time, it is proportional to the time. It is this which constitutes the principle of the conservation of areas.

WHAT IS THE CALCULUS?

The calculus is made up from two seemingly unrelated strands, now called *differentiation* and *integration*. Differentiation is concerned with how fast things move or change, and is used in the finding of velocities and tangents to curves. Integration is used to find areas of shapes in two-dimensional space or volumes in three dimensions.

As the 17th century progressed, it was gradually realized that these two strands are intimately related. As both Newton and Leibniz explained, they are inverse processes — if we follow either by the other, we return to our starting point.

However, Newton and Leibniz had different motivations, with Newton focusing on motion and Leibniz concerned with tangents and areas.

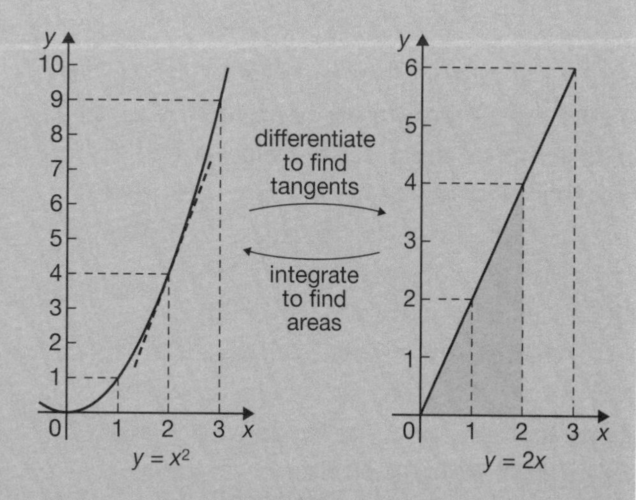

Differentiation and integration

variables and constants. A main reason for doing so was that the machinery of the calculus could then be applied both to them and to practical situations. This hastened the development of new mathematical descriptions and techniques, such as in the emerging area of *differential equations.*

This shift towards the algebraic type of description also led to a good way of discovering new objects. Books were written in the algebraic style, and mathematicians formulated, thought about and solved problems in this way. Increasingly, algebra came to be seen as a logical language suitable for the investigation of all the sciences.

Mechanics and astronomy were the main areas of practical investigation. They both applied the calculus to functions of more than one variable, such as

$$u(x, y) = x^6 + x^2y^2 + y^4;$$

here, $u(x, y)$ can be thought of as the height of a surface above the point with coordinates (x, y) in the plane.

The equations that arose were called *partial differential equations,* because they involved 'partial differentiation'. The partial derivative $\partial u/\partial x$ is the rate of change of u in the x-direction, while the partial derivative $\partial u/\partial y$ is the rate of change of u in the y-direction.

Louis XIV visits the Paris Academy of Sciences, 1671

NAPIER AND BRIGGS

In 1614 John Napier (1550–1617), 8th Laird of Merchiston (near Edinburgh), introduced logarithms as an aid to mathematical calculation, designed to replace lengthy computations involving multiplications and divisions by simpler ones using additions and subtractions. Being awkward to use, they were soon supplanted by others due to Henry Briggs (1561–1630), and their use proved an enormous boon to navigators and astronomers.

Early ideas of logarithms had appeared around the year 1500. Nicolas Chuquet and Michael Stifel listed the first few powers of 2 and noticed that to

MIRIFICI
Logarithmorum
Canonis deſcriptio,
Ejuſque uſus, in utraque
Trigonometria; ut etiam in
omni Logiſtica Mathematica,
Amplißimi, Facillimi, &
expeditiſsimi explicatio.
Authore ac Inventore,
IOANNE NEPERO,
Barone Merchiſtonii,
&c. Scoto.
EDINBURGI,
Ex officinâ ANDREÆ HART
Bibliopôlæ, CIↃ. DC. XIV.

The title page of Napier's logarithms

John Napier

multiply two of them one simply adds their exponents — so, to multiply 16 and 128 we calculate:

$16 \times 128 = 2^4 \times 2^7 = 2^{4+7} = 2^{11} = 2048,$

and write $\log_2 2048 = 11$.

NAPIER'S LOGARITHMS

The idea was not developed until Napier produced his *Mirifici Logarithmorum Canonis Descriptio* (A Description of the Admirable Table of Logarithms). This contains extensive tables of logarithms of the sines and tangents of all the angles from 0 to 90 degrees, in steps of 1 minute; Napier's use of these logarithms arose because he intended them to be used as an aid to calculation by navigators and astronomers.

Napier's logarithms are not the ones we use now. He then considered two points moving along straight lines. The first travels at constant speed for ever, while the second, representing its logarithm, moves from *P* along a finite line *PQ* in such a way that its speed at each point is proportional to the distance it still has to travel. In order to avoid the use of fractions he multiplied all his numbers by ten million.

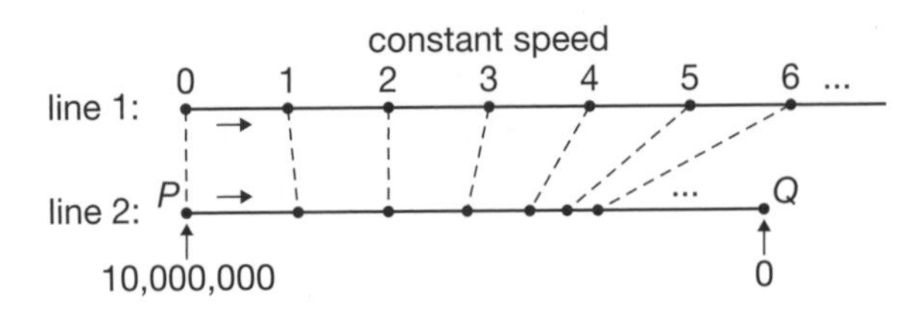

It follows from Napier's definition that the logarithm of 10,000,000 is 0. It can also be shown that, with his definition,

$\log ab = \log a + \log b - \log 1,$

for any numbers *a* and *b*; here, log 1 has the

cumbersome value of 161,180,956, which has to be subtracted in any calculation.

Napier also constructed from ivory a set of rods with numbers marked on them (now called *Napier's bones* or *rods*), which could be used to multiply numbers mechanically.

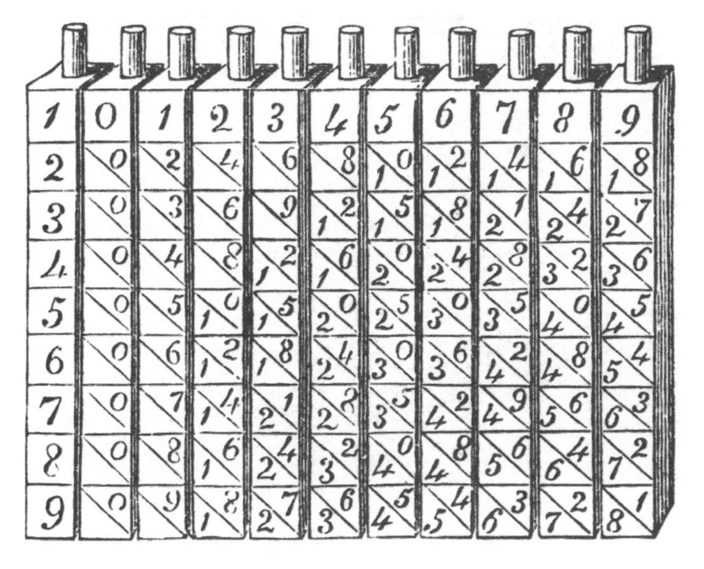

Napier's rods

HENRY BRIGGS

Shortly after their invention, Henry Briggs, first Gresham Professor of Geometry in London, heard about logarithms and enthused:

> *[John Napier] set my Head and hands a Work with his new and remarkable logarithms.*
> *I never saw a Book which pleased me better or made me more wonder.*

Briggs realized that Napier's logarithms were cumbersome, and felt that they could be redefined so as to avoid having to subtract the term log 1:

> *I myself, when expounding this doctrine to my auditors in Gresham College, remarked that it would be much more convenient that 0 should be kept for the logarithm of the whole sine.*

A related difficulty was that multiplication by 10 involved the addition of log 10 = 23,025,842.

Briggs twice visited Edinburgh to stay with Napier and sort out the difficulties. On returning to London, he devised a new form of logarithms, his logs to base 10, written $\log_{10}$, in which $\log_{10} 1 = 0$ and $\log_{10} 10 = 1$: to multiply two numbers one then simply adds their logarithms:

$$\log_{10} ab = \log_{10} a + \log_{10} b;$$

in general, if $y = 10^x$, then $\log_{10} y = x$. In 1617 he published these in a small printed pamphlet, *Logarithmorum Chilias Prima* (The First Thousand Logarithms).

In 1624, after he had left London to become the first Savilian Professor of Geometry in Oxford, Briggs followed this with an extensive collection of logarithms to base 10 of the integers from 1 to 20,000 and 90,000 to 100,000, all calculated by hand to fourteen decimal places. The gap in his tables (from 20,000 to 90,000) was soon filled in, by the Dutch mathematician Adriaan Vlacq in 1628.

2		Logarithmi.			Logarithmi.
1	0	0000,00000,00000	34	1	5314,78917,04226
2	0	3010,29995,66398	35	1	5440,68044,35028
3	0	4771,21254,71966	36	1	5563,02500,76729
4	0	6020,59991,32796	37	1	5682,01724,06700
5	0	6989,70004,33602	38	1	5797,83596,61681
6	0	7781,51250,38364	39	1	5910,64607,02650
7	0	8450,98040,01426	40	1	6020,59991,32796
8	0	9030,89986,99194	41	1	6127,83856,71974
9	0	9542,42509,43932	2	1	6232,49290,39790
10	1	0000,00000,00000	43	1	6334,68455,57959

Some of Henry Briggs's logarithms

The invention of logarithms quickly led to the development of mathematical instruments based on a logarithmic scale. Most notable among these was the *slide rule,* versions of which first appeared around 1630 and were widely used for over 300 years until the advent of the pocket calculator in the 1970s.

FERMAT

Pierre de Fermat (1601–1665) spent most of his life in Toulouse as a lawyer. He considered mathematics as a hobby, published little, and communicated with other scientists by letter. Fermat was the first important European number theorist since Greek times and resurrected the subject with some stunning results. His other main area of interest was analytic geometry, which he helped to introduce.

Fermat was born in Beaumont-de-Lomagne in the south of France and attended the University of Toulouse. After receiving a Bachelor of Civil Law degree from Orléans in 1631, he spent the rest of his life following a full-time career as a lawyer in Toulouse.

ANALYTIC GEOMETRY

Analytic geometry, in which we use algebraic techniques to help us solve geometrical problems, was born in 1637 and had two fathers: René Descartes and Pierre de Fermat. Fermat, in particular, gave a successful new method of finding tangents to curves; this used ideas and techniques of approximation taken from Diophantus's *Arithmetica*, which had recently been published in French (see opposite).

Pierre de Fermat

NUMBER THEORY

Although Fermat made substantial contributions to the development of analytic geometry, he is mainly remembered for his contributions to number theory, even though he often stated his results without proof and did not publish his conclusions.

FERMAT PRIMES

Fermat conjectured that if n is a power of 2, then $2^n + 1$ is a prime number. The first few of these numbers are indeed prime: $2^1 + 1 = 3$, $2^2 + 1 = 5$, $2^4 + 1 = 17$, $2^8 + 1 = 257$, $2^{16} + 1 = 65{,}537$. But Euler proved that $2^{32} + 1$ is divisible by 641, and no other Fermat primes have ever been found.

THE $4n + 1$ THEOREM

Let us list all prime numbers of the form $4n + 1$ (that is, each is one more than a multiple of 4):

5, 13, 17, 29, 37, 41, 53, 61, ...

Fermat observed that

Every prime number in this list can be written as the sum of two perfect squares:

for example, $13 = 4 + 9 = 2^2 + 3^2$

and $41 = 16 + 25 = 4^2 + 5^2$.

Fermat stated this result without proof. It was left for later mathematicians to prove it.

PELL'S EQUATION

Earlier we saw that Brahmagupta found integer solutions to 'Pell's equation' $Cx^2 + 1 = y^2$, for several specific values of C. His work was continued by Fermat, who managed to find a solution in the difficult case $C = 109$ and

AN ILLUSTRATION OF FERMAT'S APPROACH

Fermat used approximations to find the maximum and minimum values of certain expressions. One of his problems was to find the point E on a line AC, for which the product $AE \times EC$ takes its maximum value.

Let $AC = b$ and $AE = a$; then $EC = b - a$ and the product $AE \times EC = a \times (b - a)$.

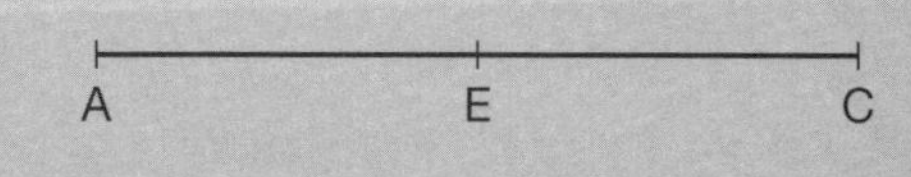

Fermat's idea was that when the position of E gives the maximum value, then this product does not change much if E is moved slightly, say by an amount e.
So $a \times (b - a)$ is approximately equal to $(a + e) \times (b - a - e)$.
When this is the case, be is approximately equal to $2ae + e^2$
and, on dividing through by e, we see that b is approximately equal to $2a + e$.
Then Fermat wrote:

Suppressing e: b = 2a. To solve the problem we must take the half of b.

So the maximum value of the product occurs when E is the midpoint of the line AC.

challenged his mathematical contemporaries to do the same. Since the smallest solution is

$x = 15{,}140{,}424{,}455{,}100$,

$y = 158{,}070{,}671{,}986{,}249$,

it is clear that Fermat must have had a general method for finding such solutions, but he never revealed it to anyone.

FERMAT'S 'LITTLE THEOREM'

Another result of Fermat concerns certain large numbers that are divisible by primes. To illustrate it, we choose a prime number, such as 37, and then choose any positive integer, such as 14. Fermat's result tells us that if we were to calculate the number $14^{37} - 14$, then the result can be divided exactly by 37. In general, Fermat's 'little theorem' tells us that:

Given any prime number p and any whole number n, the number $n^p - n$ can be divided exactly by p.

This is not simply a theoretical fact — it now forms the basis of important recent work in cryptography and internet security.

FERMAT'S 'LAST THEOREM'

We have seen that there are integers x, y and z satisfying the equation $x^2 + y^2 = z^2$ (the *Pythagorean triples*): for example, we could take $x = 3$, $y = 4$ and $z = 5$, since $3^2 + 4^2 = 5^2$.

But can we find integers x, y and z satisfying the equations $x^3 + y^3 = z^3$ and $x^4 + y^4 = z^4$, or (in general) $x^n + y^n = z^n$, for any larger number n?

In his copy of Diophantus's *Arithmetica,* Fermat claimed to have 'an admirable proof which this margin is too narrow to contain' of the statement that:

For any integer n (greater than 2), there do not exist positive integers x, y and z for which $x^n + y^n = z^n$.

Using a method he devised called the 'method of infinite descent', Fermat proved this for $n = 4$, but it seems rather unlikely that he had a general argument that works for all values of n.

As we shall see later, *Fermat's last theorem* (as it became known) was eventually proved in 1995 by Andrew Wiles, after a long and difficult struggle.

DESCARTES

René Descartes (1596–1650) was born in Touraine in France, trained as a lawyer in Poitiers, and died in Stockholm while tutor to Queen Christina of Sweden. In common with many other 17th-century thinkers, he sought a symbolism to discover truths about the world, and he saw algebraic language as a way forward that might have wider application. Cartesian coordinates are named after him.

René Descartes's ground-breaking mathematics appears in his *Discours de la Méthode* (Discourse on Method), a philosophical treatise on universal science that appeared in 1637. It had three appendices: one on optics (containing the first published statement of the law of refraction), one on metereology (containing an explanation of primary and secondary rainbows), and one on geometry, extending to 100 pages and containing fundamental contributions to analytic geometry.

La Géométrie generated great excitement and had considerable influence through its use of algebraic methods to solve geometrical problems. This set in train a gradual movement from geometry towards algebra that continued for about one hundred years, culminating in the work of Leonhard Euler.

LA GÉOMÉTRIE

Descartes first introduced a simplification by considering all quantities as non-dimensional, whereas geometers had previously dealt with lengths and considered the product of two lengths as an area.

He then claimed to be able to solve geometrical problems by algebra and he believed that solutions to algebraic equations could also be obtained by geometrical construction. As an example of such a construction he showed how to use lines and a circle to obtain a solution to a quadratic equation, as follows.

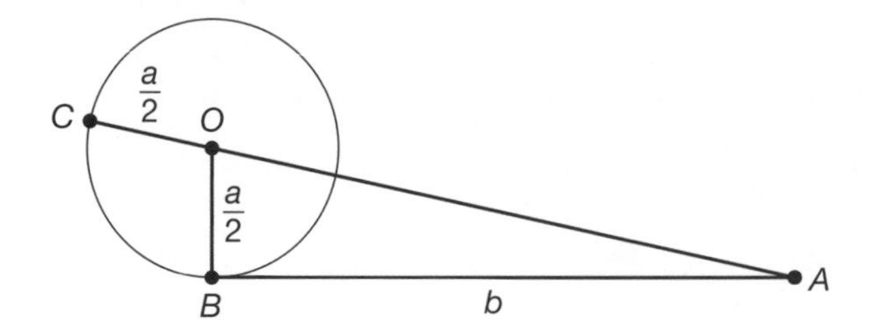

The length of AC is the positive solution of the quadratic equation $x^2 = ax + b^2$

He then illustrated his algebraic approach by solving an ancient geometrical problem of Pappus, who had asked for the path traced by a point moving in a specified way relative to a number of fixed lines. Descartes named two particular lengths x and y and then calculated all the other lengths in terms of them, thereby obtaining a equation involving the terms x^2, xy and y^2; this is a quadratic equation, and shows that the required path is a conic, as below.

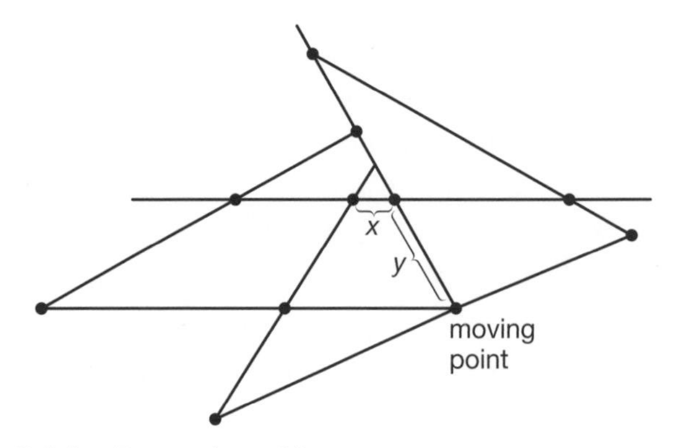

Solving Pappus's problem

Descartes also developed a method for finding tangents to curves which reduced the problem to finding solutions of a certain type of algebraic equation. However, at no stage did he

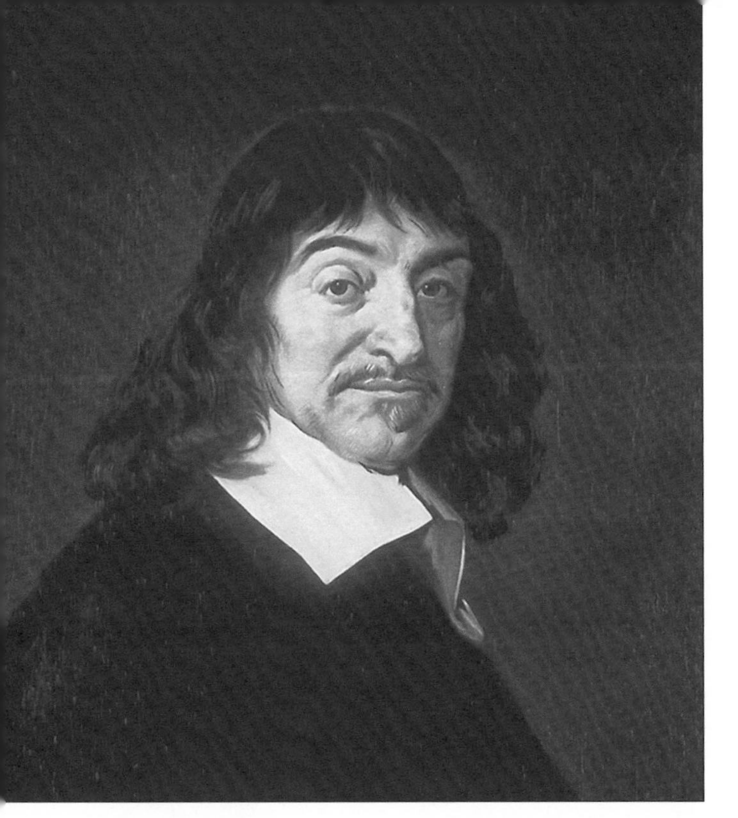

René Descartes

introduce the 'Cartesian coordinates' (with axes at right angles) that are usually named after him, with the points represented by pairs of numbers (x, y) and the lines represented by linear equations of the form $y = mx + c$.

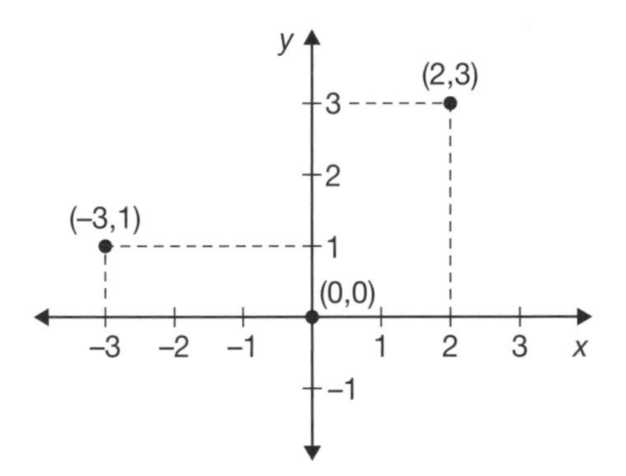

Cartesian coordinates

TWO MATHEMATICAL CONTRIBUTIONS

Descartes invented a 'rule of signs' for locating the roots of polynomial equations. Note that, as we move from left to right, the equation

$$x^4 - x^3 - 19x^2 + 49x - 30 = 0$$

has three changes of sign (between − and +, or + and −) and one pair of like signs (both −). Descartes's rule asserts that, in this situation, the equation has at most three positive solutions and at most one negative solution.

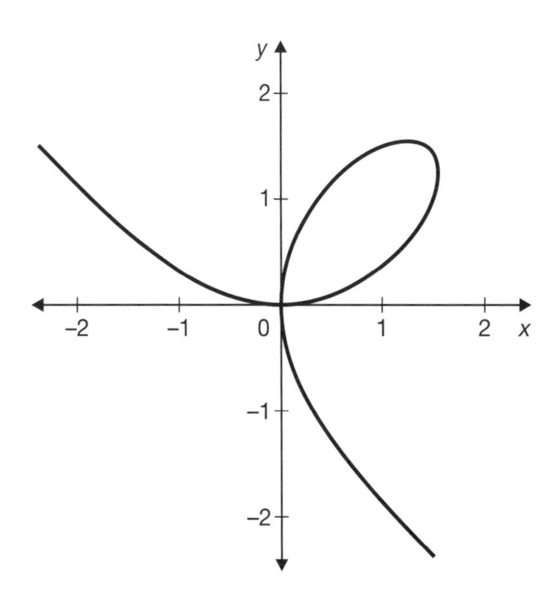

He also analysed various geometrical curves. One of these was the above *folium of Descartes,* with equation $x^3 + y^3 = 3xy$.

DESCARTES' VORTEX THEORY

Descartes also developed an influential theory of planetary motion in which vortices (whirlpools) fill space and push the planets around in their orbits (see below). This theory was later dismissed by Isaac Newton in his *Principia Mathematica.*

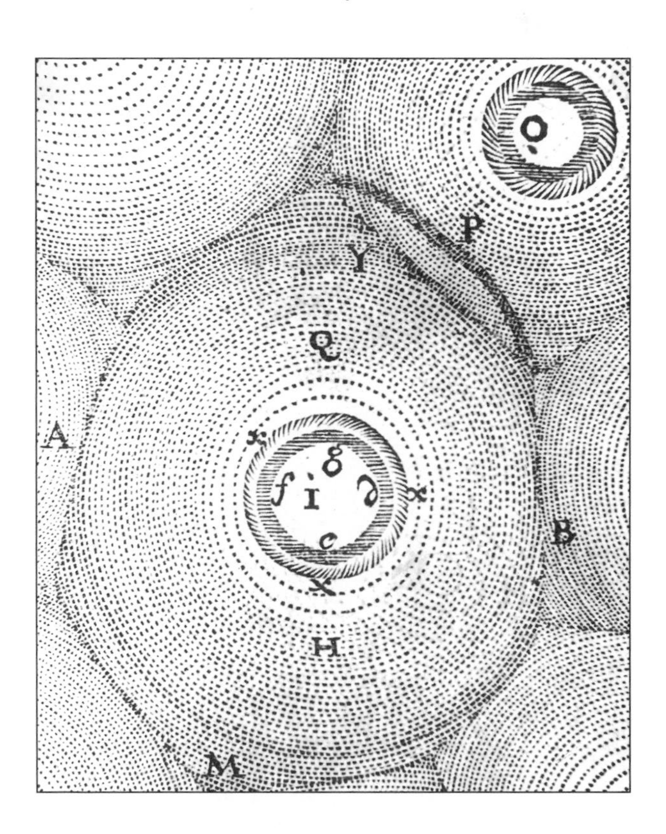

PASCAL

Blaise Pascal (1623–1662) contributed to several areas of mathematics, as well as to science and religious philosophy. He laid the foundations for the theory of probability, wrote about atmospheric pressure, discovered his 'hexagon theorem' in projective geometry, and built a calculating machine. He also wrote an influential treatise on the array of numbers now known as 'Pascal's triangle'.

Blaise Pascal was born in Clermont-Ferrand in the Auverne region of France, and showed his mathematical ability at an early age. His father Étienne, a tax official, lawyer and amateur mathematician, took responsibility for his son's education and brought him to the scientific meetings in Paris arranged by Marin Mersenne.

PASCAL'S CALCULATING MACHINE

In 1642 Pascal built a calculating machine (known as 'The Pascaline') to help in his father's work. Although it could only add or subtract, using 'carry-over' gearing, it was well documented — indeed, many later machines were only slight modifications of it.

Blaise Pascal

PROBABILITY

The modern theory of probability is often considered to begin in 1654, arising out of some correspondence between Pascal and Fermat over gambling problems. One particular problem was raised by the Chevalier of Méré and concerns the fair division of the stakes in a game when it is interrupted before its conclusion.

Suppose that two players agree to play a certain game repeatedly; the winner, who wins £100, is the one who first wins six times. If the game is interrupted when one player has won five games and the other player has won four games, how should the £100 be divided fairly between the players? (The answer is that the one with five wins receives £75 and the other receives £25.) Pascal gave the general solution in more detail in his *Traité du Triangle Arithmétique* (Treatise on the Arithmetical Triangle) of 1654.

The Pascaline

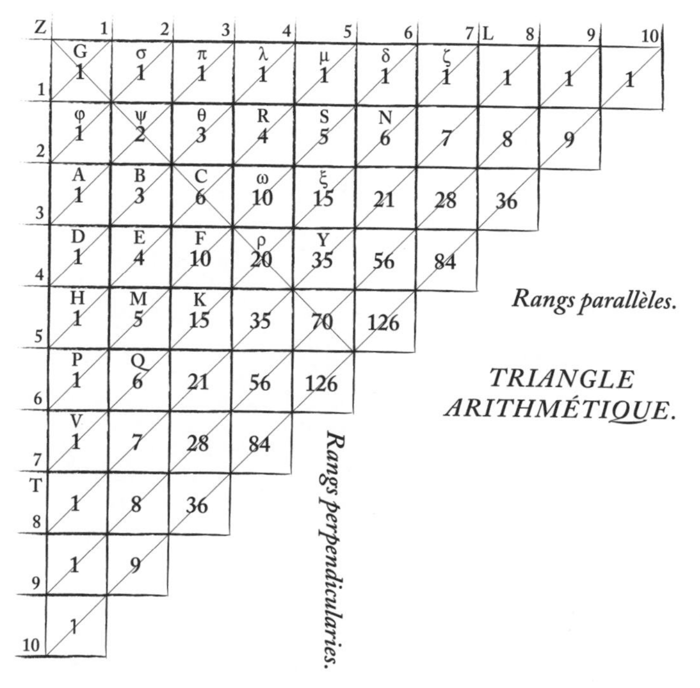

Pascal's drawing of his triangle (adapted from his posthumous publication of 1665)

THE ARITHMETICAL TRIANGLE

The arithmetical triangle, now called *Pascal's triangle,* was known to Islamic, Indian and Chinese mathematicians many centuries earlier. But to credit Pascal is justifiable because he carried out the first systematic investigation into its properties.

Over the centuries the numbers in this array arose in various different ways.

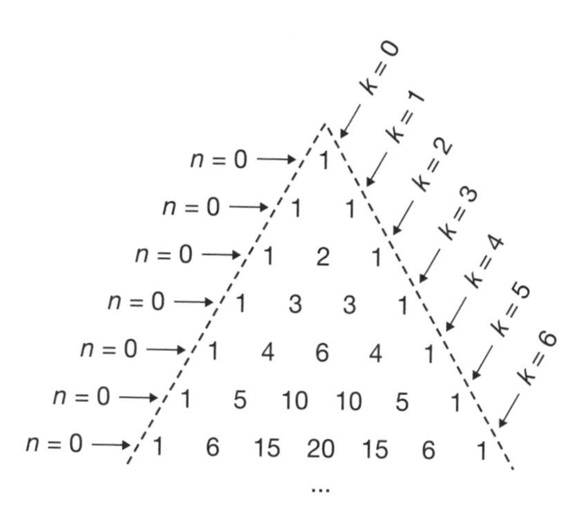

The first few rows of Pascal's triangle

PATTERNS OF NUMBERS

The arithmetical triangle has several interesting numerical features. For example, the first diagonal ($k = 0$) is a sequence of 1s, and the next two diagonals ($k = 1$ and 2) contain the natural numbers 1, 2, 3, ... and the triangular numbers 1, 3, 6, 10, ...

Moreover, each number (apart from the outside 1s) is the sum of the two numbers above it; for example, the number 20 in the seventh row ($n = 6$) is the sum of the two 10s above it.

Another interesting feature is that the sum of the entries in each row is a power of 2; for example, in the sixth row ($n = 5$):

$$1 + 5 + 10 + 10 + 5 + 1 = 32 = 2^5.$$

Pascal proved this result using a method, now called *mathematical induction,* which he was the first to express explicitly.

BINOMIAL COEFFICIENTS

The numbers appearing in the triangle are all *binomial coefficients.* These numbers arise in the expansions of the various powers of $1 + x$:

$$(1 + x)^0 = 1$$
$$(1 + x)^1 = 1 + 1x$$
$$(1 + x)^2 = 1 + 2x + 1x^2$$
$$(1 + x)^3 = 1 + 3x + 3x^2 + 1x^3$$
$$(1 + x)^4 = 1 + 4x + 6x^2 + 4x^3 + 1x^4$$
$$(1 + x)^5 = 1 + 5x + 10x^2 + 10x^3 + 5x^4 + 1x^5$$
$$(1 + x)^6 = 1 + 6x + 15x^2 + 20x^3 + 15x^4 + 6x^5 + 1x^6$$

COMBINATIONS

The numbers in the triangle also arise as the number of different ways of making selections; for example, the number of different 4-member teams that can be chosen from 6 people is written $C(6, 4)$ and is 15. In general, $C(n, k)$ is the number in row n and diagonal k, and is equal to $n! / k! (n - k)!$.

CAVALIERI AND ROBERVAL

During the 17th century, much progress was made on the two branches of the infinitesimal calculus, the seemingly unrelated areas now called *differentiation* and *integration.* Bonaventura Cavalieri (1598–1647) developed a theory of 'indivisibles' that provided a systematic way to help in the calculation of certain areas. Gilles Personne de Roberval (1602–1675) was a French scientist who also discovered powerful techniques for calculating areas.

Cavalieri, an Italian mathematician, was held in high regard by Galileo, who declared:

Few, if any, since Archimedes, have delved as far and as deep into the science of geometry.

In 1629, Galileo helped Cavalieri to obtain a professorship at Bologna which was renewed every three years until his death.

Roberval at the Paris Academy of Sciences, 1666

CAVALIERI'S PRINCIPLE

Cavalieri wrote ten books on mathematics and science and published a table of logarithms. His most important work was his *Geometria Indivisibilibus Continuorum Nova Quadam Ratione Promota* (A Certain Method for the Development of a New Geometry of Continuous Indivisibles), which appeared in 1635.

Cavalieri considered a geometrical object to be made up of objects of one dimension lower, the *indivisibles,* so an area was made up from lines and a solid object was made up from planes. The problem was then how to compare the indivisibles of one geometrical object with those of another.

Cavalieri's principle set out circumstances in which this can be done. In the case of areas it says that:

Two plane figures have the same area if they lie between the same parallel lines, and any line drawn parallel to the given two lines cuts off equal chords in each figure.

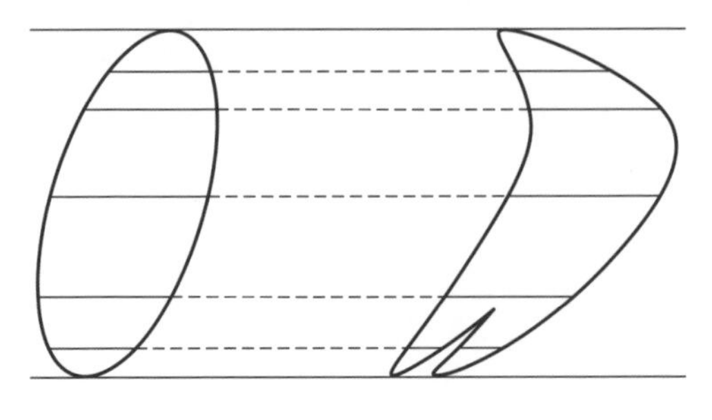

Cavalieri's principle for plane figures

THE ROUTE TO THE CALCULUS

Many other 17th-century mathematicians, such as Kepler, Fermat, Descartes and Pascal, found tangents to curves and areas under curves. There were also contributions from:

- **Gregory of St Vincent, a Belgian mathematician who found the area under the hyperbola $y = x^{-1}$.**
- **John Wallis, who found the area under the curve $y = x^k$, when k is a positive fraction.**
- **Evangelista Torricelli, a student of Galileo and the inventor of the barometer, who found areas and tangents and studied the parabolic paths of projectiles. In going from an equation for 'distance in terms of time' to one of 'speed in terms of time', and conversely, he became aware of the inverse nature of tangent and area problems.**
- **Isaac Barrow (Newton's predecessor as Lucasian Professor of Mathematics in Cambridge), who also studied the inverse relationship between these problems.**

However, it was Newton and Leibniz who transcended what anyone had done before, and independently created what today we call the calculus:

- ***differentiation* — a systematic way of obtaining slopes of tangents to curves**
- ***integration* — a systematic way of obtaining areas under curves**
- **the *inverse relationship between these tangent and area problems* — namely, that differentiation and integration are inverse processes: if you integrate an expression and then differentiate the result, you get back to where you started, and *vice versa*.**

Cavalieri used this principle to find the area under the curve $y = x^n$, where n is a given positive integer.

ROBERVAL AND THE CYCLOID

Cavalieri's principle was widely viewed as useful and powerful. Roberval made impressive use of it (claiming to have discovered it independently) to calculate the area under one arch of a *cycloid,* the curve traced by a fixed point on a circle rolling along a straight line; one can think of a cycloid as the curve traced out by a piece of mud on a bicycle tyre when the bicycle is wheeled along.

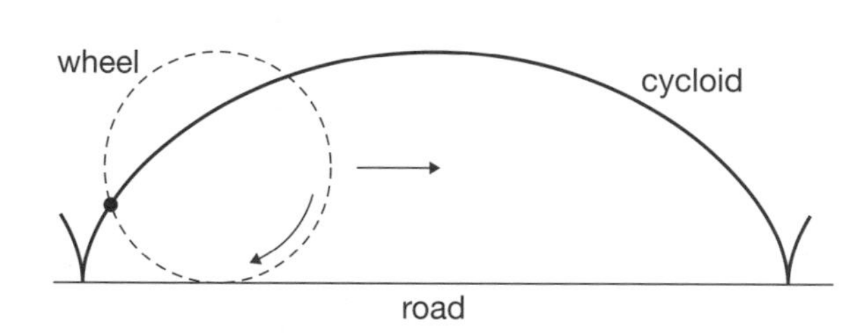

A cycloid

Roberval proved that the area under one arch of a cycloid is exactly three times that of the area of the generating circle. To do this, he showed that the two horizontally shaded regions below have the same area, and noted that the vertically shaded area is one-half that of the rectangle *OABC.* He deduced that the area under one-half of the cycloid is

$\frac{1}{2}\pi r^2 + \frac{1}{2}(2r \times \pi r) = \frac{3}{2}\pi r^2$.

It follows that the total area under one arch of the cycloid is $3\pi r^2$, which is three times the area of the circle.

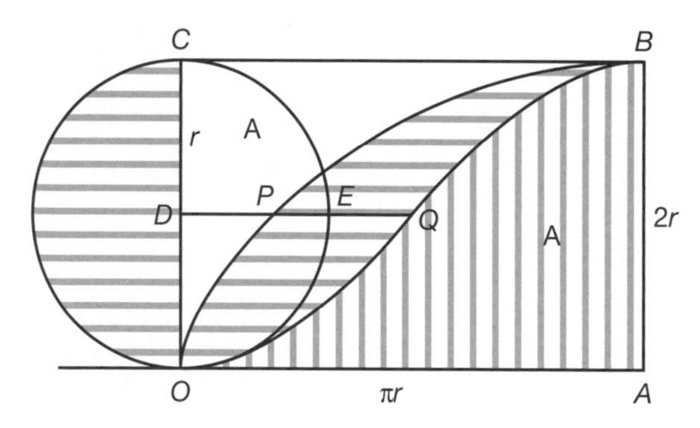

Roberval's cycloid

HUYGENS

The 1650s saw a shift in the centre of mathematical activity away from France. The Netherlands and Britain were now the leading countries, and foremost among Dutch mathematicians was Christiaan Huygens (1629–1695). With his construction of the first pendulum clock, Huygens considerably improved the accuracy of time-measurement. He also contributed to geometry, mechanics, astronomy and probability.

Huygens belonged to a prominent and well-connected Dutch family. He studied law and mathematics, first at the University of Leiden and then at the College of Orange at Breda. His first works in the 1650s were on mathematics, studying the cissoid and conchoids (classical curves studied by the Greeks).

PENDULUM CLOCKS

The need for clocks to measure time accurately was important in astronomy and navigation. A recurring interest of Huygens was the development of such clocks, and his most famous discovery was in this area.

Christiaan Huygens

For a pendulum bob, the period of oscillation is only approximately independent of the amplitude (or extent) of the oscillation, although this inaccuracy is often ignored when the amplitude is small. Huygens discovered that:

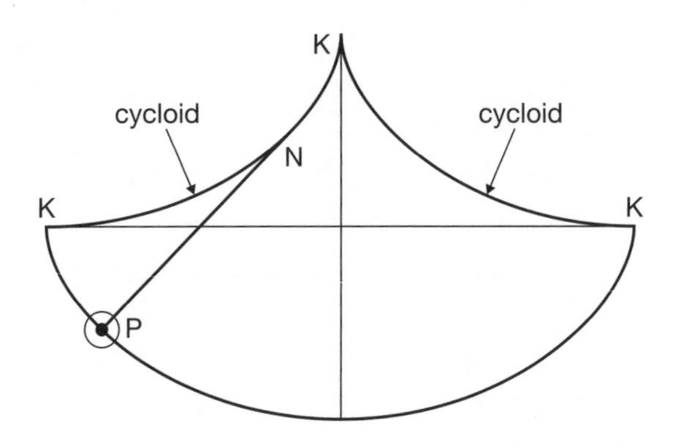

A pendulum bob following a cycloidal path

For a cycloid whose axis is erected on the perpendicular and whose vertex is located at the bottom, the times of descent, in which a body arrives at the lowest point at the vertex after having departed from any point on the cycloid, are equal to each other ...

This means that, if the path of a pendulum could be made to follow a cycloidal path, its period of oscillation would be independent of its amplitude.

Huygens achieved this by having two 'cheeks' from the point of suspension of the pendulum along which the cord wraps itself as the pendulum swings. The shapes of these cheeks are also cycloids; their precise form depends on the length of the pendulum.

PROBABILITY

In 1655 Huygens became interested in probability during a visit to Paris and, encouraged by Pascal, wrote *De Ratiociniis in Ludo Aleae* (On the Values in Games of Chance), published in 1657. This was the first systematic treatment of probability theory, and remained the only one available until the 18th century. As Huygens remarked:

Although in a pure game of chance the results are uncertain, the chance that one player has to win or lose depends on a determined value.

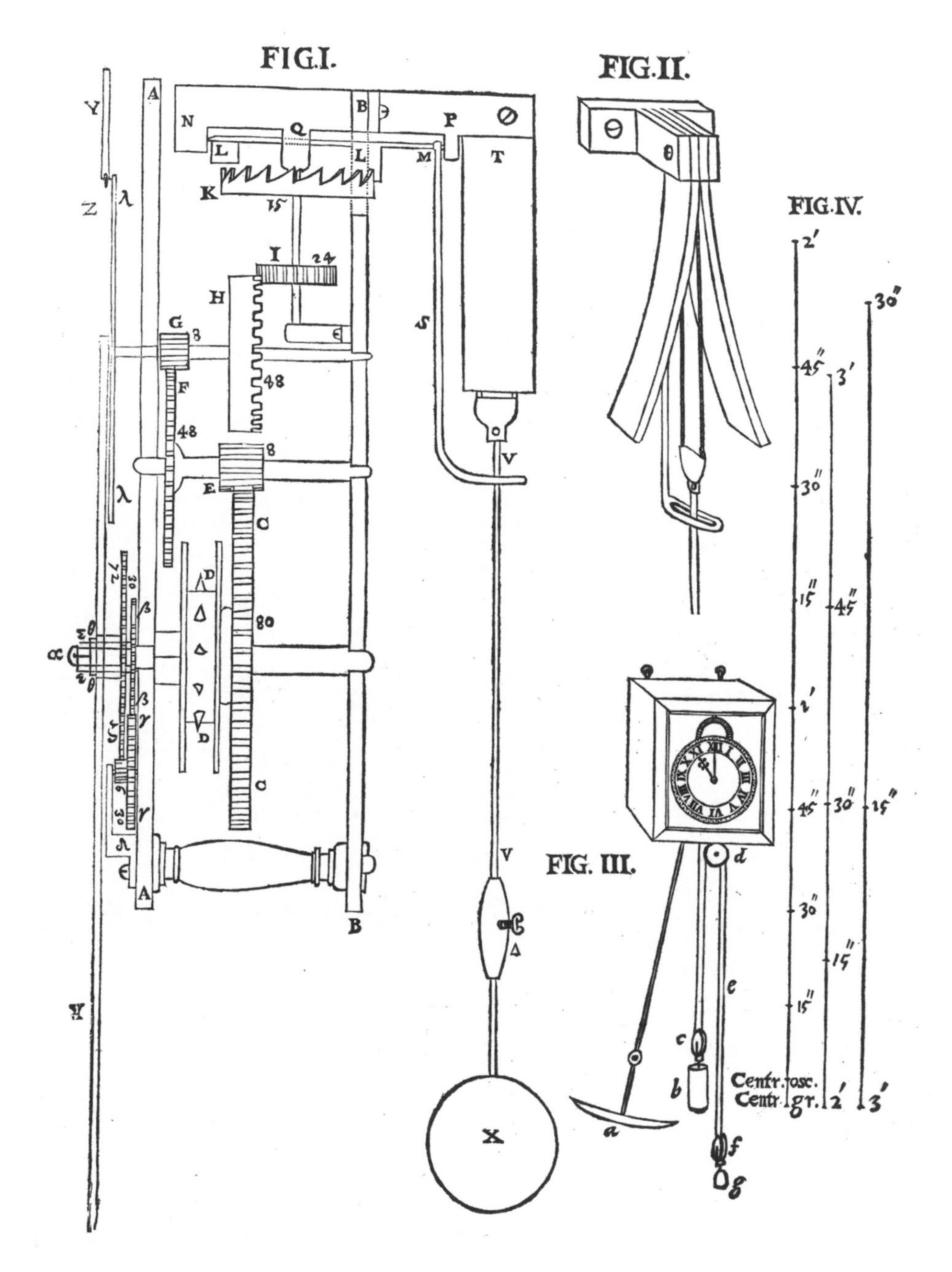

Huygens' drawings of clock mechanisms, from his *Horologium Oscillatorium* (The Pendulum Clock) of 1673

His 'determined value' is what we now call the *expectation*, the average expected winnings if the game were played many times.

An illustration of his first principle is that if in a game one has an equal chance of winning £x or £y, then the expected winnings would be £$\frac{1}{2}(x + y)$. This is the stake that a player should be prepared to bet to play the game.

Huygens also discussed a problem raised in correspondence between Pascal and Fermat:

How many times should a pair of dice be thrown so as to give at least an even chance of obtaining a double-six?

The answer lies between 24 and 25: after 24 throws the probability of obtaining a double-six is slightly less than $\frac{1}{2}$; after 25 throws it is slightly greater than $\frac{1}{2}$.

ASTRONOMY AND MOTION

In cooperation with his brother, Huygens developed great expertise in the grinding and polishing of lenses. This led to them making the best telescopes of the time, which enabled Huygens to discover, in 1655, the moon of Saturn now called Titan. In the following year he presented his description of the rings of Saturn:

It is surrounded by a thin flat ring, nowhere touching and inclined to the ecliptic.

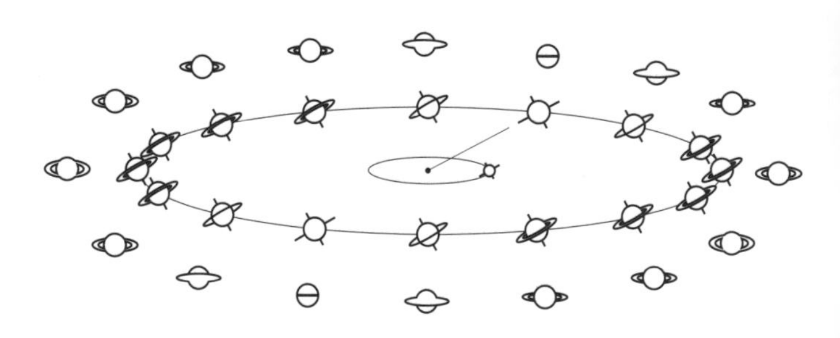

A diagram from Huygens' *Systema Saturnalia* (1659), showing the orbit of Saturn

WALLIS

John Wallis (1616–1703) was the most influential English mathematician before the rise of Newton. His most important works were his *Arithmetic of Infinites* and his treatise on conic sections, both published in the 1650s. These were full of fresh discoveries and insights and appeared at a critical time in the development of the subject. It was through studying the former that Newton came to discover his version of the binomial theorem.

John Wallis

John Wallis was born at Kent (England) and went to Cambridge University where he studied very little mathematics. His aim was to follow a church career, which he did until 1649 when he was appointed Savilian Professor of Geometry at Oxford. Prior to this he had gained little mathematical experience, and his Oxford appointment may have been due mainly to his distinguished work in deciphering secret codes for Cromwell's intelligence service during the English Civil War. A more far-sighted mathematical appointment on flimsier evidence seems hard to imagine.

PARS PRIMA.

PROP. I.

De Figuris planis juxta Indivisibilium methodum considerandis.

Suppono in limine (juxta Bonaventuræ Cavallerii *Geometriam Indivisibilium*) Planum quodlibet quasi ex infinitis lineis parallelis conflari : Vel potius (quod ego mallem) ex infinitis Parallelogrammis æque altis; quorum quidem singulorum altitudo sit totius altitudinis $\frac{1}{\infty}$, sive aliquota pars infinite parva; (esto enim ∞ nota numeri infiniti;) adeoque omnium simul altitudo æqualis altitudini figuræ.

Est item (propter tangentem) $DT \geq DO$ (hoc est, DT æqualis vel major quam DO; illud quidem si D, P, coincidant; hoc, si secus) & $DTq \geq DOq$, hoc est $\frac{f^2 \pm 2fa + a^2}{f^2} p^2 \geq \frac{d \pm a}{d} p^2$; & (utrumque multiplicando in $df2$ &

The first appearance of ∞ and ≥, in Wallis's treatise on conic sections

MATHEMATICAL LANGUAGE

Although he was often conservative in his use of mathematical notation, he did introduce two new symbols that are still in current use: ∞ for 'infinity' and ≥ for 'greater than or equal to'.

In his 1655 book on conic sections, Wallis treated conics as curves defined by equations, rather than as sections of a cone, and obtained their properties by the techniques of algebraic analysis introduced by Descartes.

Although Newton was the first to display the then-current notation for fractional and negative indices, in 1676, Wallis went far towards laying the groundwork – for example, in his *Arithmetica Infinitorum* he wrote:

$1/x$, whose index is -1
and $\sqrt{x}$, whose index is $\frac{1}{2}$.

THE AREA UNDER A PARABOLA

This area was known at the time, but Wallis went on to use his method to other cases for which the answer was unknown.

Wallis's approach relies on the relationship between the sums of different series. To illustrate his method, we find the area under the parabola $y = x^2$ by comparing it to that of a square on the same interval. We think of both areas as being composed of infinitely many vertical lines.

An approximation to the area of the parabola is the sum of the lengths of the lines above the points 0, $1/n$, $2/n$, $3/n$, ... , n/n; this sum is

$$0^2/n^2 + 1^2/n^2 + 2^2/n^2 + 3^2/n^2 + \ldots + n^2/n^2.$$

The sum of the lengths of the lines in the square above the same points is

$$1 + 1 + 1 + \ldots + 1 \quad (n + 1 \text{ terms}),$$

since each line has length 1.

The ratio of these two sums can be calculated, and is $\frac{1}{3} + \frac{1}{6n}$. This gives the correct answer of $\frac{1}{3}$ when n becomes large.

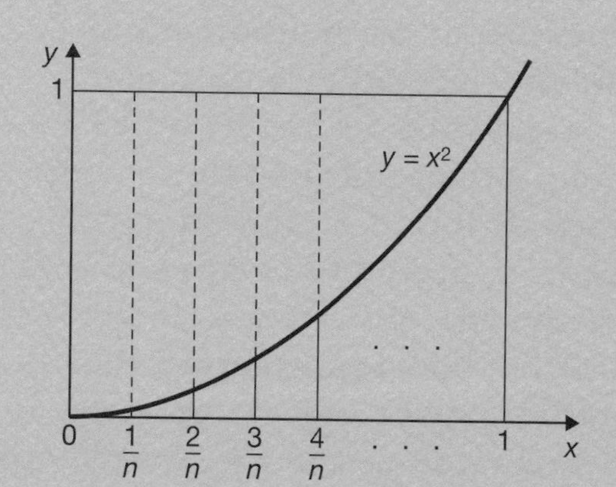

ARITHMETICA INFINITORUM

After his appointment, Wallis studied all the major mathematical works in Oxford's libraries. In particular, he encountered the works of Torricelli, in which Cavalieri's method of indivisibles was used, and felt that this method could be used to find the area of the circle. This occupied him for over three years and led to his *Arithmetica Infinitorum* (Arithmetic of Infinites), which appeared in 1655. The word *interpolation* was introduced by Wallis in this work.

Cavalieri had used the method of geometrical indivisibles to find the area under curves of the form $y = x^n$, where n is a positive integer. Wallis extended similar techniques to the curves $y = x^k$, where k is a fraction.

Johannis Wallisii, SS. Th. D.
GEOMETRIÆ PROFESSORIS
SAVILIANI in Celeberrimâ
Academia OXONIENSI,

ARITHMETICA
INFINITORVM,

SIVE

Nova Methodus Inquirendi in Curvilineorum Quadraturam, aliaq; difficiliora Matheseos Problemata.

OXONII,
Typis LEON: LICHFIELD Academiæ Typographi,
Impensis THO. ROBINSON. *Anno* 1656.

Wallis gave a full account of his methods in order, he said, to throw open the very fount to his readers, rather than to imitate the methods of the ancients who strove to be admired rather than understood, and whose methods were 'harsh and difficult, so that few would venture to approach them'.

It was here that Wallis obtained his famous formula for the ratio of the areas of a square and an inscribed circle, which we now write as $4/\pi$:

$$\frac{4}{\pi} = \frac{3 \times 3 \times 5 \times 5 \times 7 \times 7 \times \ldots}{2 \times 4 \times 4 \times 6 \times 6 \times 8 \times \ldots}$$

Newton was attracted to Wallis's fundamental method of discovery, the exploration and recognition of pattern. Wallis's career had been set in motion by his cryptological skills, and they seem also to have characterized his mathematical style.

NEWTON

Sir Isaac Newton (1642–1727) remains unchallenged in the depth and breadth of his mathematical and scientific work. He obtained the general form of the binomial theorem, explained the relationship between differentiation and integration, studied power series, and analysed cubic curves. In gravitation, Newton asserted that the force that causes objects on earth to fall is the same as the one that keeps the planets orbiting around the sun, and that they are governed by a universal inverse-square law of force.

Isaac Newton was born on Christmas Day 1642 in the hamlet of Woolsthorpe, near Grantham in Lincolnshire. He went to Cambridge University where he was later appointed Lucasian Professor at the age of 26, staying in post until 1696 when he moved to London to become Warden, and then Master, of the Royal Mint. He became President of the Royal Society in 1703.

NEWTON'S MATHEMATICAL WORKS

CALCULUS

Newton's creation of the calculus was involved with movement — how things change with time, or 'flow'. His tangent problems involved velocities, and in his *Treatise on Fluxions* (circulated to his friends, but not published until after his death) he gave rules for calculating these velocities. For area problems he did not use a direct approach, but regarded them as inverse problems.

Isaac Newton

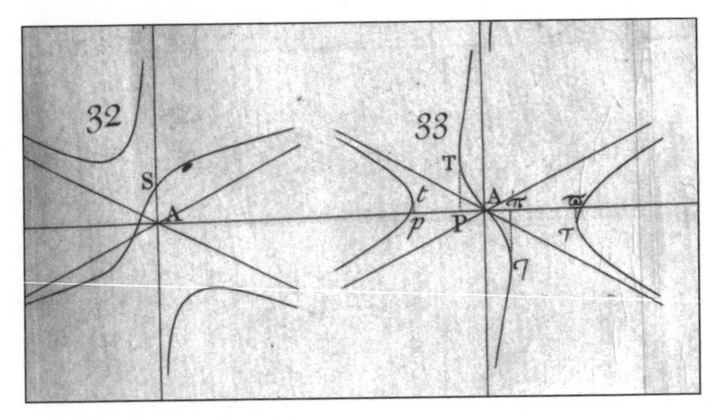

Two of Newton's cubic curves

INFINITE SERIES

An infinite series is similar to a polynomial, except that it goes on for ever — for example,

$$1 - 2x + 3x^2 - 4x^3 + 5x^4 - 6x^5 + \dots .$$

Newton's realization of the importance and use of infinite series was a major contribution to mathematics. Particularly useful was his general binomial theorem which gives the expansion of $(1 + x)^n$ as an infinite series when n is not a positive integer — for example, the infinite series above is the binomial expansion of $(1 + x)^{-2}$.

CUBIC CURVES

The classification of quadratic curves was well known — they are the conic sections — but the classification of cubic curves was much more difficult. Newton achieved it by showing that all 78 different types of them can be obtained by projection from one of a family of five.

PRINCIPIA MATHEMATICA

The celebrated story of the apple, which Newton recounted in old age, is part of scientific folklore. Seeing an apple fall, he apparently suggested that the gravitational force that pulls the apple to earth is the same as the force that keeps the moon orbiting around the earth and the earth orbiting around the sun. Moreover, this planetary motion is

PHILOSOPHIÆ
NATURALIS
PRINCIPIA
MATHEMATICA.

Autore *J. S. NEWTON*, *Trin. Coll. Cantab. Soc.* Mathefeos Profeffore *Lucafiano*, & Societatis Regalis Sodali.

IMPRIMATUR.
S. PEPYS, *Reg. Soc.* PRÆSES.
Julii 5. 1686.

LONDINI,

Juffu *Societatis Regiæ* ac Typis *Jofephi Streater*. Proftat apud plures Bibliopolas. *Anno* MDCLXXXVII.

governed by a universal law of gravitation, the *inverse-square law:*

> *The force of attraction between two objects varies as the product of their masses, and inversely as the square of the distance between them*

– so if each mass is trebled then the force increases by a factor of 9, and if the distance between them is multiplied by 10 then the force decreases by a factor of 100.

In his *Philosophiae Naturalis Principia Mathematica* (Mathematical Principles of Natural Philosophy) (1687), possibly the greatest scientific work of all time, Newton used this law and his three laws of motion to explain Kepler's three laws of elliptical planetary motion and to account for the orbits of comets, the variation of tides, and the flattening of the earth at its poles due to the earth's rotation.

In his *Principia* Newton also considered the movement of objects in resisting media and the velocity needed to project an object so as to put it into orbit around the earth.

NEWTON'S THREE LAWS OF MOTION

- Every body continues in its state of rest, or of uniform motion in a right line, unless it is compelled to change that state by forces impressed on it.
- Any change of motion is proportional to the force, and is made in the direction of the line in which the force is applied.
- To any action there is an equal and opposite reaction.

Newton achieved honour and reverence in his own lifetime. Alexander Pope's well-known epitaph expresses the reverence with which Newton's contemporaries viewed him:

> *Nature, and Nature's Laws lay hid in Night.*
> *God said, Let Newton be! And All was light.*

Newton's memorial in Westminster Abbey, London

WREN, HOOKE AND HALLEY

Christopher Wren (1632–1723), Robert Hooke (1635–1703) and Edmond Halley (1656–1742) all contributed to the early history of London's Royal Society. All three played significant roles in the mathematical life of late 17th-century England, although none of them was primarily a mathematician.

Bust of Christopher Wren

CHRISTOPHER WREN

Although Wren is mainly remembered as an architect, his early career was as a distinguished astronomer, while Isaac Newton ranked him (with Wallis and Huygens) as one of the outstanding geometers of the age.

Wren entered Wadham College, Oxford, in 1646, and impressed his contemporaries with his youthful ability. In the 1650s he was elected a Fellow of All Souls College, where the fine sundial he designed still stands. He became a regular member of the Oxford Philosophical Society, a group of brilliant colleagues (including Wallis, Hooke and Boyle) who met regularly to discuss topics of scientific interest and perform experiments.

In 1657 Wren was appointed Professor of Astronomy at London's Gresham College. In his inaugural lecture he enthused about the London mathematical scene, and concluded:

> *Mathematical demonstrations being built upon the impregnable Foundations of Geometry and Arithmetick are the only truths that can sink into the Mind of Man, void of all Uncertainty; and all other discourses participate more or less of Truth according as their Subjects are more or less capable of Mathematical Demonstration.*

On 28 November 1660, following one of Wren's astronomy lectures at Gresham College, the assembled company met in the Geometry professor's rooms and proposed that a new society be founded to promote experimental science. Two years later this became the Royal Society.

Wren's interest in geometry also pervaded his architecture. For the flat roof structure of his Sheldonian Theatre in Oxford, Wren considered a model that had been designed some years earlier by John Wallis. Designing the interlocking of the beams involved the formulation and solution of twenty-seven simultaneous algebraic equations.

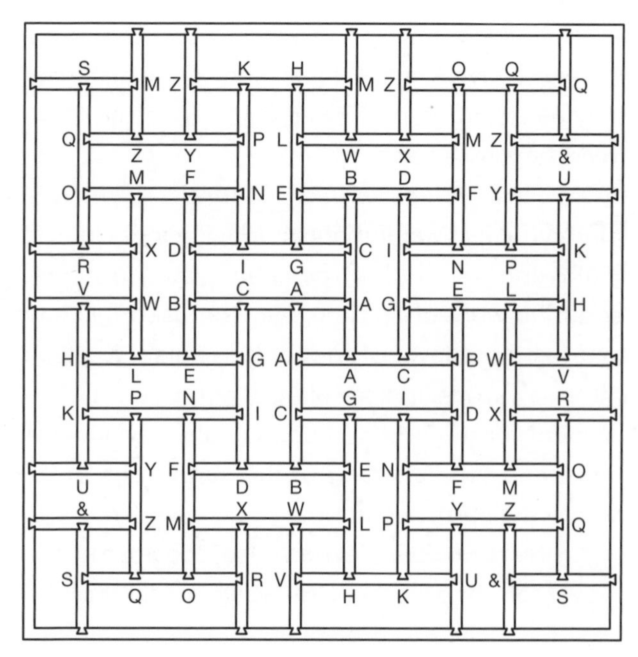

A model for the Sheldonian roof

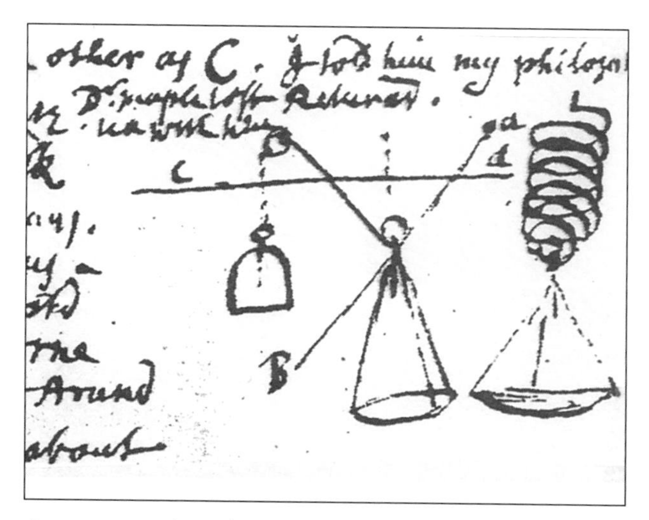

A note on springs from Hooke's diary

ROBERT HOOKE

For over thirty-five years Hooke was professor of geometry at Gresham College, giving lectures on mathematics to the general public. The Royal Society held its meetings at the College and Hooke, as the Society's Curator of Experiments, was required to design and present experiments on a regular basis. In this way, the College became an important centre for scientific research and debate.

Hooke was interested in the mathematical principles underlying many of his experiments, and designed a number of mathematical instruments. He was also interested in the design of clocks and watches and formulated 'Hooke's law for springs': if a weight is attached to a spring, the resulting extension of the spring is proportional to the weight added.

EDMOND HALLEY

In 1684, at the Royal Society, Wren, Hooke and Halley were trying to decide the orbit of a planet moving under the influence of a central inverse-square law of gravitation. Halley travelled to Cambridge to ask Newton:

> *Sr Isaac replied immediately that it would be an Ellipsis, the Doctor struck with joy & amazement asked him how he knew it, why saith he I have calculated it, whereupon Dr Halley asked him for his calculation without any further delay, Sr Isaac looked among his papers but could not find it, but he promised him to renew it, & then to send it to him.*

As a result of this conversation, Halley spent the next three years cajoling and persuading Newton to rework his calculations and present them for publication. The result was Newton's *Principia Mathematica,* which opens with Halley's fulsome *Ode to Newton:*

> *Then ye who now on heavenly nectar fare,*
> *Come celebrate with me the name*
> *Of Newton, to the Muses dear; for he*
> *Unlocked the hidden treasuries of truth ...*
> *Nearer the gods no mortal may approach.*

Without Halley the *Principia* would never have appeared. Moreover, as the Royal Society were unable to pay for its publication, having spent its money on a lavish *History of Fishes,* Halley paid for it himself; the Society rewarded him with fifty copies of the *History of Fishes.*

The founding of the Royal Society

In 1704, following the death of John Wallis, Halley was appointed Oxford University's Savilian Professor of Geometry. He prepared a definitive edition of Apollonius's *Conics,* and in 1720 became the second Astronomer Royal. Basing his calculations on Newtonian principles, he accurately predicted the return of a comet in December 1758; it was subsequently named *Halley's comet.*

LEIBNIZ

Leibniz's calculating machine

Gottfried Wilhelm Leibniz (1646–1716) was the greatest theorist of logic and language since Aristotle, and a mathematician and philosopher of the highest rank. He was guided in his investigations by the desire for a 'logic of discovery' and a language that would reflect the structure of the world. This showed itself in his work on binary arithmetic, symbolic logic, his calculus and his calculating machine.

Leibniz was born in Leipzig and went to the University there at the early age of 14 (where his father was professor of moral philosophy), and after that to Altdorf, where he took his doctoral degree while still only 20. He was exceptionally talented, with wide-ranging interests across many disciplines, but did not obtain an academic position after leaving university. He spent forty years in rather minor positions, travelling around Europe and representing the interests of the Elector of Mainz and the Duke of Hanover.

BINARY ARITHMETIC

Following ideas of Ramón Lull, Leibniz set out his guiding inspiration in an essay of 1666: the intention to devise

> *A general method in which all truths of the reason would be reduced to a kind of calculation.*

His binary arithmetic was an example of the attempt to reduce complex ideas to their simplest forms, and around 1679 he wrote that, instead of the decimal system:

> *It is possible to use in its place a binary system, so that as soon as we have reached two we start again from unity in this way:*
>
> (0) (1) (2) (3) (4) (5) (6) (7) (8)
>
> 0 1 10 11 100 101 110 111 1000
>
> *... what a wonderful way all numbers are expressed by unity and nothing.*

Such binary representations are now routinely used by modern computers.

Gottfried Leibniz

THE PRIORITY DISPUTE

Who invented the calculus first?

Newton was probably the first to discover his results on the calculus, but although he circulated his discoveries privately to his friends, they were not published until after his death.

Leibniz, working independently, introduced his superior notation in 1675 and published his results on the differential calculus in 1684 and on the integral calculus in 1686. In the latter paper, he also explained the inverse relationship between differentiation and integration.

This led to a bitter priority dispute between Newton and Leibniz, with Newton's followers accusing Leibniz of plagiarism. With much ill-feeling between Britain and the Continent on this issue, Newton (as President of the Royal Society) arranged for an 'independent' commission to investigate the issue. It was not Newton's finest hour: he personally chose the members of the commission, writing much of the evidence for them to consider, and the commission unsurprisingly ruled in his favour.

A CALCULATING MACHINE

Leibniz's calculating machine shows another aspect of his plan, in which he wished to use mechanical calculation to find a way to error-free truth. The machine's crucial innovation was a stepped gearing wheel with a variable number of teeth along its length, which allowed multiplication on turning a handle. Leibniz's stepped wheel was an important component of mechanical calculators until they were replaced by electronic calculators.

THE CALCULUS

Leibniz's calculus was by far his most ambitious and influential work, and again arose from his desire to find general symbolic methods for uncovering truths.

Leibniz's calculus originated in a different manner from that of Newton's, being based on sums and differences, rather than on velocity and motion.

In 1675 Leibniz introduced two symbols that would forever be used in calculus. One was his d (or dy/dx) notation for differentiation, referring to a decrease in dimension — for example, from areas (x^2) to lengths (x). The other was the integral sign: attempting to find areas under curves by summing lines, he defined *omnia l* (all the *ls*), which he then represented by an elongated S for sum: this is the symbol $\int$ for the integral sign.

It will be useful to write $\int$ *for omn...*

Leibniz presented algebraic rules for differentiation which can then be used to find tangents, and to locate maxima and minima. As well as giving the rules, he also differentiated powers of x:

$d(x^a) = a\, x^{a-1} dx$, where a is any fraction

— so $d(x^2) = 2x\, dx$ and $d(x^{1/2}) = \frac{1}{2}\, x^{-1/2}\, dx$.

LEIBNIZ'S RULES FOR DIFFERENTIATION

- For any constant a:
  $d(a) = 0,\ d(ax) = a\, dx$
- $d(v + y) = dv + dy$
- $d(vy) = v\, dy + y\, dv$
- $d(v/y) = (y\, dv - v\, dy) / y^2$

These rules are easy to use — for example, we can use them to differentiate

$w = x^{1/2}/(x^2 + 4)$.

In the last rule, let $v = x^{1/2}$ and $y = x^2 + 4$.

Then

$dw = \{(x^2 + 4)\, d(x^{1/2}) - x^{1/2}\, d(x^2 + 4)\} / (x^2 + 4)^2$.

By the second rule,

$d(x^2 + 4) = d(x^2) + d(4) = d(x^2) + d(4) = d(x^2)$,

since $d(4) = 0$, by the first rule.

Finally, on substituting

$d(x^2) = 2x\, dx$ and $d(x^{1/2}) = \frac{1}{2}\, x^{-1/2}\, dx$,

we get:

$dw = ((x^2 + 4).\frac{1}{2}x^{-1/2}\, dx - x^{1/2}.2x\, dx) / (x^2 + 4)^2$,

which can be rearranged as

$dw = \{(2 - \frac{3}{2}x^2) / x^{1/2}.(x^2 + 4)^2\}\, dx$.

JACOB BERNOULLI

In the entire history of science and mathematics it is hard to find a more prominent family than the Bernoullis. Its first distinguished member was Jacob Bernoulli (1654–1705), who was born in Basel in Switzerland and later became professor of mathematics there in 1687. He had a wide range of interests, investigating infinite series, the cycloid, transcendental curves, the logarithmic spiral and the catenary, and introduced the term *integral*. His posthumous text on probability contains the celebrated law of large numbers.

Jacob Bernoulli remained as Professor of Mathematics at Basel until his death, when he was succeeded by his brother Johann. The Bernoulli brothers were the foremost advocates of the Leibnizian calculus, publicizing it, publishing on it, and applying it to solve new problems.

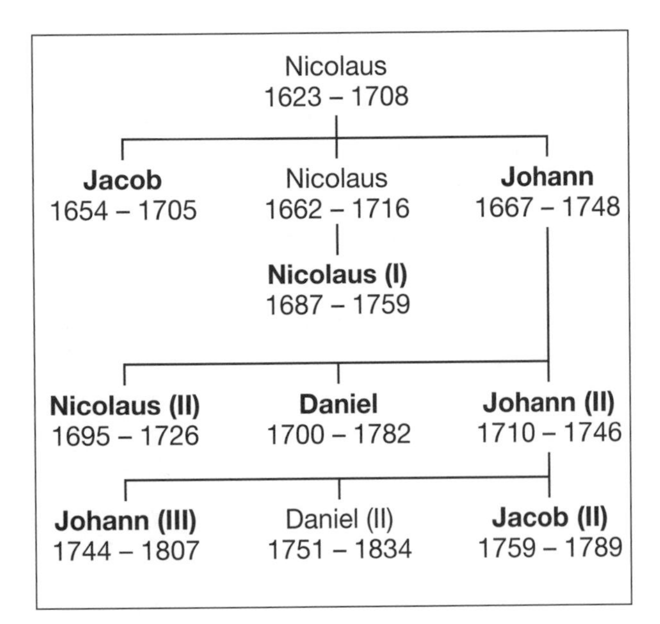

The Bernoulli mathematical dynasty, with eight mathematicians in bold type

Jacob confirmed that the harmonic series

$$1 + \tfrac{1}{2} + \tfrac{1}{3} + \tfrac{1}{4} + \tfrac{1}{5} + \tfrac{1}{6} + \ldots$$

does not converge to a finite number, while the series

$$1 + (\tfrac{1}{2})^2 + (\tfrac{1}{3})^2 + (\tfrac{1}{4})^2 + (\tfrac{1}{5})^2 + (\tfrac{1}{6})^2 + \ldots$$

converges — but he could not find its sum.

The brothers Jacob and Johann competed fiercely and publicly. On one occasion, Jacob posed the problem of finding the shape that a heavy chain takes up when hung between two points. Galileo had incorrectly thought it to be a parabola, and Johann was delighted to beat his brother to the correct answer, which is a curve that we call a *catenary*.

A catenary

ARS CONJECTANDI

Jacob Bernoulli's book on probability, *Ars Conjectandi* (Art of Conjecturing), was his most important and influential publication. It was the culmination of twenty years work and was published in 1713, eight years after his death.

JACOBI BERNOULLI,
Profess. Basil. & utriusque Societ. Reg. Scientiar. Gall. & Pruss. Sodal.
MATHEMATICI CELEBERRIMI,
ARS CONJECTANDI,
OPUS POSTHUMUM.
Accedit
TRACTATUS
DE SERIEBUS INFINITIS,
Et EPISTOLA Gallicè scripta
DE LUDO PILÆ
RETICULARIS.

BASILEÆ,
Impensis THURNISIORUM, Fratrum.
cIↄ Iↄcc xiii.

The first three parts of the book build on previous work – indeed, the first part is a commentary on Huygens' earlier treatment. Here, he calculated the sums of integer powers (squares, cubes, ...), obtaining a general result involving what are now called the *Bernoulli numbers.*

The last part was very innovative. His central concern was how to quantify probabilities in situations where it was impossible to list or count all possibilities. His approach was to see what happened in similar situations:

> *For example, if we have observed that out of 300 persons of the same age and with the same constitution as a certain Titius, 200 died within ten years while the rest survived, we can with reasonable certainty conclude that there are twice as many chances that Titius will have to pay his debt to nature within the ensuing decade as there are chances that he will live beyond that time.*

Bernoulli believed that the more observations we make, the better we can predict future outcomes, and he quantified this in his *law of large numbers.* He showed that increases in the number of observations enable us to estimate the probability to any degree of accuracy, and he calculated how many observations are needed so that we are sure to be within a predefined degree of accuracy. To do this involved working with sums of binomial coefficients.

THE LOGARITHMIC SPIRAL

Bernoulli was interested in the logarithmic spiral and called it *spira mirabilis* (the marvellous spiral). It has the property that the tangent at each point makes a fixed angle with the line joining the point to the centre. This spiral has pleasing symmetries, reproducing itself under various transformations – for example, each arm of the spiral has the same shape as the previous one, but is larger.

Bernoulli asked for this spiral to be carved on his tombstone, with the inscription *EADEM MUTATA RESURGO* (I arise the same, though changed). It appears at the bottom of the tombstone (see right).

JOHANN BERNOULLI

Johann Bernoulli (1667–1748), the younger brother of Jacob, was a prolific mathematician who tutored the Marquis de l'Hôpital and Leonhard Euler. He posed many problems in connection with the priority debate between Newton and Leibniz, and of these, the most significant for the development of mathematics was that of the *brachistochrone* — finding the curve of quickest descent. Bernoulli was called the 'Archimedes of his age', and this is inscribed on his tombstone.

Johann Bernoulli was born and died in Basel. He developed Leibniz's techniques in the 1690s and with Huygens' support occupied the chair of mathematics at Groningen (Holland) from 1695 until 1705, when he succeeded his brother in Basel.

Johann Bernoulli

THE CALCULUS

Before moving to Groningen Bernoulli was employed in France as tutor to the Marquis de l'Hôpital in the Leibnizian calculus. This resulted in l'Hôpital's *Analyse des Infiniment Petits, pour l'Intelligence des Lignes Courbes* (The Analysis of the Infinitely Small, for the Understanding of Curved Lines) (1696), the first printed book on the differential calculus. It contained many of Johann's results, including a result for calculating limits now usually known as *l'Hôpital's rule.* The Marquis credited Bernoulli in his preface:

> I *must own myself very much obliged to the labours of Messieurs Bernoulli, but particularly to those of the present Professor at Groeningen, as having made free with their Discoveries as well as those of Mr Leibniz: so that whatever they please to claim as their own I frankly return them.*

Johann, however, considered that this had been done in a rather offhand manner.

Bernoulli's comprehensive work on the integral calculus was published in 1742, although most of it had been written by 1700. He defined integration as the inverse of differentiation and gave several techniques for evaluating integrals. He explained that its main use was in finding areas, and then used it to solve *inverse tangent problems,* in which we are given some property of the tangent to a curve at each point and are asked to find the curve. Most importantly, Bernoulli showed how to recast problems in geometry or mechanics in the language of the calculus; this caused inverse tangent problems to become known as *differential equations,* because they were recast as equations involving differentials.

THE BRACHISTOCHRONE PROBLEM

In June 1696 Bernoulli proposed the following 'new problem which mathematicians are invited to solve', concerning an object descending under gravity along a curve from a point A to a point B:

If two points A and B are given in a vertical plane, to assign to a mobile particle M the path AMB along which, descending under its own weight, it passes from the point A to the point B in the briefest time.

At first, one might think that the 'curve of quickest descent' from A to B is the straight line joining them, but this is not the case – and nor do we achieve the shortest time if the curve is too steep to start with and too flat later on. The desired compromise between these is known as the *brachistochrone,* from the Greek words for 'shortest time'.

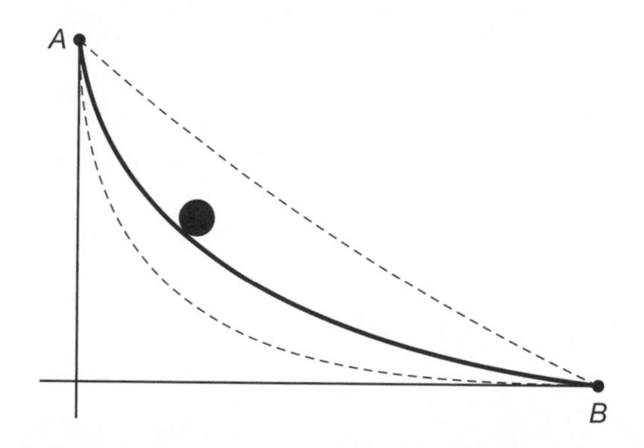

Using an optical analogy, Bernoulli deduced that at each point on the desired curve, the sine of the angle between the tangent to the curve and the vertical axis is proportional to the square root of the distance fallen. This gave rise to the differential equation

$dy/dx = \sqrt{x} / (1 - x)$,

which he then solved to show that the desired curve is a *cycloid,* the curve that Roberval had studied and that Huygens used in constructing his pendulum clock.

The brachistochrone problem was answered by Jacob Bernoulli, Leibniz and Newton. Jacob Bernoulli showed, using the calculus, that no matter where an object starts on this curve, it descends to the bottom after the same amount of time. Newton solved the problem overnight and sent his solution anonymously, but on seeing it, Johann Bernoulli identified Newton's style, saying that 'I recognize the lion by his claw'.

The brachistochrone problem gave rise to new strands of mathematics. Jacob Bernoulli's method for solving it used an approach that initiated a whole new field, the *calculus of variations,* in which one seeks a curve that satisfies a given maximum or minimum property – here, the cycloid minimizes the time of descent.

Meanwhile, arising out of his investigations into the curve of quickest descent, Johann posed the problem of finding two families of curves with the property that each curve in the first family meets each curve in the second family at right angles. These are called *orthogonal families,* and led to new concepts in thinking about expressions that depend on more than one variable.

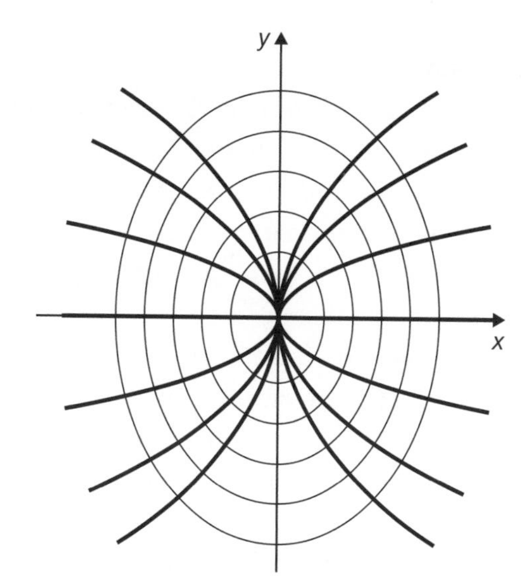

Two orthogonal families of curves

NEWTON'S SUCCESSORS

The appearance of Newton's *Principia* in 1687 caused a sensation – but scientists were puzzled by the nature of an attractive force of gravity that could apparently act over astronomical distances: to Huygens, in particular, it was an 'absurd' idea that was not capable of explaining anything. Preferable to them was some sort of mechanical theory such as that of Descartes, in which the planets are swept along by vortices like leaves in a whirlpool. But there were two main areas in which Newton's theory caused difficulties – the shape of the earth and the motion of the moon. Both were important for navigation, and in both of them Newton's views were eventually vindicated.

In 1728, the year after Newton's death, the great French author, historian and philosopher Voltaire wrote about the different world views in France and England:

> *In Paris they see the universe as composed of vortices of subtle matter; in London they see nothing of the kind. For us it is the pressure of the moon that causes the tides of the sea; for the English it is the sea that gravitates towards the moon.*

Voltaire was well placed to comment, as he had the expertise of Madame du Châtelet to inform him. She was the gifted mathematician who translated Newton's *Principia* into French. Voltaire continued:

> *In Paris you see the earth shaped like a lemon; in London it is flattened on two sides.*

A melon (oblate spheroid) and a lemon (prolate spheroid)

THE SHAPE OF THE EARTH

Under Newton's hypotheses, the rotation of the earth causes a flattening at the poles so that the earth is melon-shaped, whereas under Descartes's vortex and matter theory there is an elongation at the poles, so that the earth is lemon-shaped.

To decide on the actual shape of the earth, the Paris Academy sent two expeditions to measure

Emilie de Breteuil, Marquise du Châtelet

the size of a degree of latitude: one to Peru, led by Charles-Marie de la Condamine in 1735, and the other to Lapland, headed by Pierre de Maupertuis in 1736. It was not until 1739 that both expeditions reported, and Maupertuis was able to confirm that Newton was right: the earth is flatter at the poles. This earned Maupertuis the nickname of 'the great flattener'.

Although this vindicated Newton's approach, it turned out that Newton had incorrectly calculated the amount of flattening because of assumptions he had made about fluid pressure, although he correctly predicted the nature of the earth's shape.

Alexis Claude Clairaut

THE MOTION OF THE MOON

Although Newton dealt well with the motion of two bodies moving under mutual gravitational attraction, the motion of the moon depends not only on the earth but also on the sun. Even today we have no exact solution of the *three-body problem* — the problem of predicting the future positions and speeds of three bodies moving under mutual gravitational attraction.

Without the influence of the sun the motion of the moon would be an ellipse. Newton simplified the problem by assuming that the effect of the sun was to cause the moon's elliptical orbit to revolve slowly. He calculated that it would take eighteen years for the orbit to return to its original position, but observation showed that it took only nine years. As Newton wrote in later editions of the *Principia*:

The apse of the moon is about twice as swift.

By the end of the 1740s Newton's theory of gravitation was under concerted investigation by those mathematicians who best understood it: d'Alembert, Clairaut and Euler. In 1747 Clairaut, who had taken part in Maupertuis's Lapland expedition, proposed to modify Newton's inverse-square law of gravitation by adding an additional term to it, while d'Alembert and Euler came up with other approaches. It seemed that Newton's law of gravitation might be wrong!

Then, on 17 May 1749, Clairaut made a dramatic retraction:

> *I have been led to reconcile observations on the motion of the moon with the theory of attraction without supposing any other attractive force than one proportional to the inverse square of the distance.*

Clairaut had taken a new approach to the differential equations that describe the moon's motion, finding that the previous differences between theory and observation had been due to the way in which these equations were approximated.

This led to Euler publishing his theory of the moon in 1753, which enabled the astronomer Tobias Mayer to prepare a set of tables describing its motion — enabling the moon to be used as a 'celestial clock'. This led eventually to their receiving a share of the prize awarded by the British Board of Longitude for discovering a practical way of finding longitude at sea.

D'ALEMBERT

Jean le Rond d'Alembert (1717–1783) was a leading Enlightenment figure. In his later years he wrote many of the mathematical and scientific articles for Denis Diderot's celebrated *Encyclopédie,* which attempted to classify the knowledge of the time. Earlier, he had been the first to obtain the wave equation that describes the motion of a vibrating string. He also attempted to formalize the idea of a limit so as to put the calculus on a firm basis, and studied the convergence of infinite series, obtaining a result now known as the ratio test.

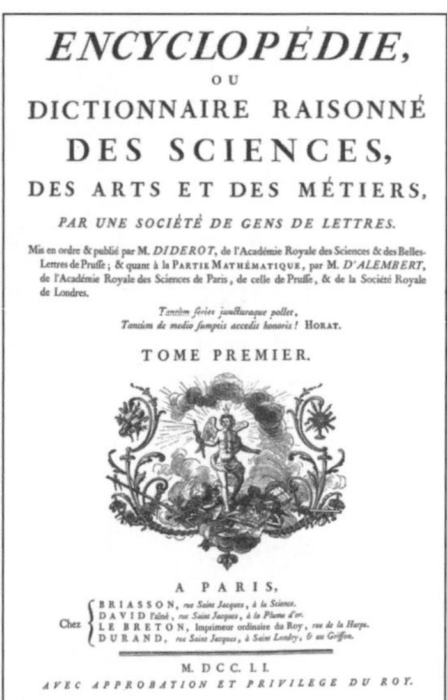

ENCYCLOPÉDIE,
OU
DICTIONNAIRE RAISONNÉ
DES SCIENCES,
DES ARTS ET DES MÉTIERS,
PAR UNE SOCIÉTÉ DE GENS DE LETTRES.

Mis en ordre & publié par M. *DIDEROT*, de l'Académie Royale des Sciences & des Belles-Lettres de Pruſſe ; & quant à la PARTIE MATHÉMATIQUE, par M. *D'ALEMBERT*, de l'Académie Royale des Sciences de Paris, de celle de Pruſſe, & de la Société Royale de Londres.

Tantùm ſeries juncturaque pollet,
Tantùm de medio ſumptis accedit honoris! HORAT.

TOME PREMIER.

A PARIS,
Chez BRIASSON, *rue Saint Jacques, à la Science.*
DAVID l'aîné, *rue Saint Jacques, à la Plume d'or.*
LE BRETON, Imprimeur ordinaire du Roy, *rue de la Harpe.*
DURAND, *rue Saint Jacques, à Saint Landry, & au Griffon.*

M. DCC. LI.
AVEC APPROBATION ET PRIVILEGE DU ROY.

The first volume of the *Encyclopédie*

As an infant, d'Alembert was abandoned by his mother outside the church of St Jean le Rond near Notre Dame in Paris — hence his name — and was brought up by the wife of a glazier. Although he qualified as a lawyer in 1738, his main interest was in mathematics. He was apparently a brilliant conversationalist and gifted with a superb memory. He was also quick to quarrel — with Clairault, Euler and Daniel Bernoulli, among others.

Jean le Rond d'Alembert

THE *ENCYCLOPÉDIE*

The *Encyclopédie* was published between 1751 and 1777, with contributions from over 140 people, and consisted of over 70,000 articles. It was the major achievement of the French Enlightenment whose aim, in Diderot's words, was to 'change the common way of thinking'. D'Alembert was a leading member of the group of philosophers — *A Society of People of Letters,* as the title page shows — who produced this bible of the Enlightenment.

THE VIBRATING STRING

D'Alembert made notable contributions to the analysis of the motion of a vibrating string. If a string is stretched horizontally between two fixed points and made to vibrate, then, as he noted in a paper of 1747, the vertical displacement $u(x, t)$ of the string depends on both the horizontal distance x and the time t.

D'Alembert's contribution was to obtain a differential equation that describes the motion of the string. This was the first time that the

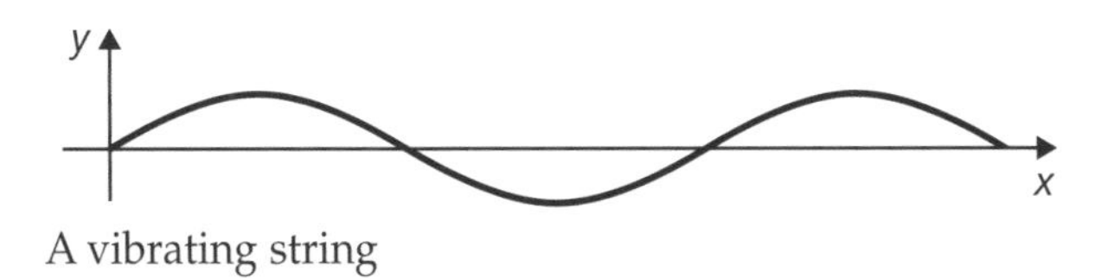

A vibrating string

techniques of the calculus had been deployed on a problem with more than one variable, and involves differentiation with respect to both x and t. The differential equation that he discovered is now called the *wave equation:*

$$c^2\, \partial^2 u(x, t)/\partial x^2 = \partial^2 u(x, t)/\partial t^2,$$

where c is a constant that depends on the string.

D'Alembert solved this 'partial differential equation' to find the motion of the string. His solutions were very general, but this is hardly surprising, since the string can be released from any initial shape with any initial velocity. His solution was

$$u(x, t) = f(x + ct) + g(x - ct),$$

where f and g are arbitrary functions. As he remarked:

This equation contains an infinity of curves.

The question of how general such solution curves can be became one of the most stimulating mathematical questions of the 18th century.

In 1752 d'Alembert tried to find solutions of the form

a function of time × a function of distance

— that is, $u(x, t) = F(t) \times G(x)$, where F depends on t and G depends on v. This converts the wave equation in two independent variables into two differential equations, each in one variable. These are easier to solve, and he deduced the solution

$$u(x, t) = \cos\,(k\pi c/L)t \times \sin\,(v\pi/L)x,$$

where L is the length of the string and k can be any positive integer. Different values of k give different modes of vibration with different frequencies.

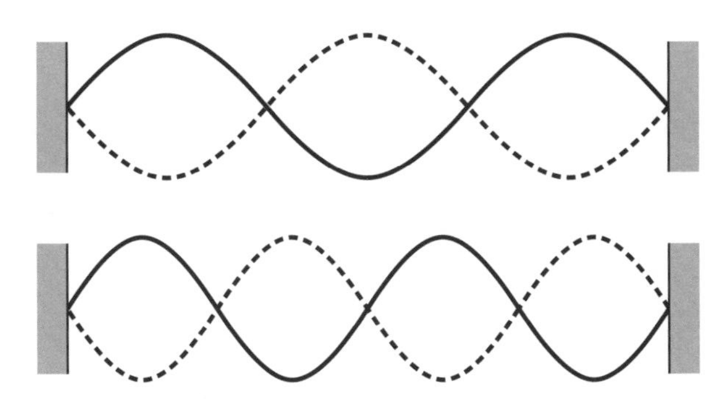

Two modes of vibration (k = 3 and 4)

It was Daniel Bernoulli, the son of Johann, who suggested that a vibrating string can exhibit infinitely many modes of vibration, all superimposed on each other. The solution is then the infinite sum:

$$u(x, t) = \alpha \cos\,(\pi c/L)t \sin\,(\pi/L)x + \beta \cos\,(2\pi c/L)t \sin\,(2\pi/L)x + \gamma \cos\,(3\pi c/L)t \sin\,(3\pi/L)x + \ldots$$

D'ALEMBERT AND THE IDEA OF A LIMIT

In his 1734 book *The Analyst*, Bishop Berkeley of Cloyne (in Ireland) had severely criticized the calculus of Newton and Leibniz for its shaky foundations. D'Alembert was concerned about such criticisms, and attempted to rescue the situation by basing the calculus on the idea of a 'limit'. In one of his contributions to the *Encyclopédie* he wrote:

One says that one quantity is the limit of another quantity, when the second can approach the first more closely than any given quantity, however small, without the quantity approaching, passing the quantity which it approaches; so that the difference between a quantity and its limit is absolutely inassignable.

D'Alembert was only partially successful in his quest, and it was not until 1821 that the task was finally accomplished, by Augustin-Louis Cauchy.

EULER

Leonhard Euler (1707–1783) was the most prolific mathematician of all time. He produced over eight hundred books and papers in a wide range of areas, from such 'pure' topics as number theory and the geometry of a circle, via mechanics, logarithms, infinite series and calculus, to such practical concerns as optics, astronomy and the stability of ships. He also introduced the symbols *e* for the exponential number, *f* for a function and *i* for $\sqrt{-1}$. In the words of Laplace: *Read Euler, read Euler, he is the master of us all.*

Euler's life can be conveniently divided into four periods. He spent his early years in Basel, Switzerland, entering the University there at the age of 14 and receiving personal instruction from Johann Bernoulli. At the age of 20 he moved to Peter the Great's newly founded St Petersburg Academy, where he became head of the mathematics division. From 1741 to 1766 he was at Frederick the Great's Academy of Sciences in Berlin, before returning to St Petersburg for his final years.

EULER'S BOOKS

Leonhard Euler wrote several groundbreaking books. His *Introductio in Analysin Infinitorum* (Introduction to the Analysis of Infinite Quantities) of 1748 expounded on infinite series, the exponential function, the properties of conics, partitions of numbers, and much else besides.

In 1755, he published a massive tome on the differential calculus, reformulating the subject in terms of the idea of a function and containing all the latest results, many due to him. He followed this, in 1768, with an influential three-volume work on the integral calculus, and in 1772 with a 775-page account of the motion of the moon.

His best-known work, still in print today, was his *Letters to a German Princess,* written from Berlin to the Princess of Anhalt-Dessau on a range of scientific topics.

Leonhard Euler

EULER, THE POLYMATH

We now survey just a few of the myriad contributions that Euler made to mathematics.

INFINITE SERIES

While at St Petersburg, Euler became interested in infinite series. We have seen that the 'harmonic series' of reciprocals has no finite sum, but Euler noticed that adding the first n terms of this series (up to $1/n$) gives a value very close to $\log_e n$. In fact, as he demonstrated, the difference between them,

$$(1 + {}^1\!/_2 + {}^1\!/_3 + {}^1\!/_4 + {}^1\!/_5 + \ldots + {}^1\!/_n) - \log_e n,$$

tends to a limiting value close to 0.577, now called *Euler's constant.*

A difficult problem of the time, known as the *Basel problem,* was to find the sum of the reciprocals of the perfect squares:

$$1 + {}^1\!/_4 + {}^1\!/_9 + {}^1\!/_{16} + {}^1\!/_{25} + \ldots ;$$

the answer was known to be about 1.645, but no-one could find its exact value. Euler achieved

THE EXPONENTIAL FUNCTION

We have all heard of 'exponential growth', meaning something that grows very fast. Such growth arises in connection with compound interest or population growth, while there is 'exponential decay' in the decay of radium or the cooling of a cup of tea.

Expressions such as 2^n or 3^n grow much faster than n^2 or n^3 as n increases; for example, if $n = 50$, then a computer calculating a million numbers per second can count up to $n^3 = 125{,}000$ in $^1/_8$ of a second, but would take 23 billion years to count up to 3^n.

In fact, mathematicians usually consider, not 2^n or 3^n, but e^n, where $e = 2.6182818\ldots$. The reason for choosing this strange number e is that if we plot the curve $y = e^x$, then

***the slope of this curve at any point x is also e^x* — that is, $dy/dx = y$ for each point of the curve.**
Such a simple differential equation holds for $y = e^x$ and its multiples, but for no other curves.

The exponential function e^x turns up throughout mathematics and its applications — for example, Euler wrote it as a limit: e^x is the limit of $(1 + x/n)^n$, as x becomes large,

and expanded as an infinite series: $e^x = 1 + x/1! + x^2/2! + x^3/3! + \ldots$;

in particular, $e = 1 + {}^1/_{1!} + {}^1/_{2!} + {}^1/_{3!} + \ldots$.

Also, the exponential function is the inverse of the logarithm function: *if $y = e^x$, then $x = \log_e y$*.

Euler's most celebrated achievement was to extend the above infinite series to complex numbers, obtaining the result $e^{ix} = \cos x + i \sin x$, which intriguingly links the exponential function with the trigonometrical ones. A special case of this, relating the most important constants in mathematics, is $e^{i\pi} = -1$ or $e^{i\pi} + 1 = 0$.

fame by showing that the sum is $\pi^2/6$. He then extended his calculations, ingeniously finding the sum of the reciprocals of all the 4th powers ($\pi^4/90$), the 6th powers ($\pi^6/945$), and so on, right up to the 26th powers!

MECHANICS

Throughout his life, Euler was interested in mechanics. In 1736, he published *Mechanica*, a 500-page treatise on the dynamics of a particle. Later, in work on the motion of rigid bodies, he obtained what we now call *Euler's equations of motion* and coined the phrase *moment of inertia.* Further results on mechanics were obtained in the 1770s. Much of this work used differential equations, an area to which Euler contributed a great deal.

THE BRIDGES OF KÖNIGSBERG

In 1735 Euler solved a well-known recreational problem. The city of Königsberg in East Prussia consisted of four regions joined by seven bridges, and its citizens used to entertain themselves by trying to cross each bridge exactly once. Can this be done?

Using a counting argument, involving the number of bridges emerging from each region, Euler proved that such a walk is impossible. He then extended his arguments to any arrangement of land areas and bridges.

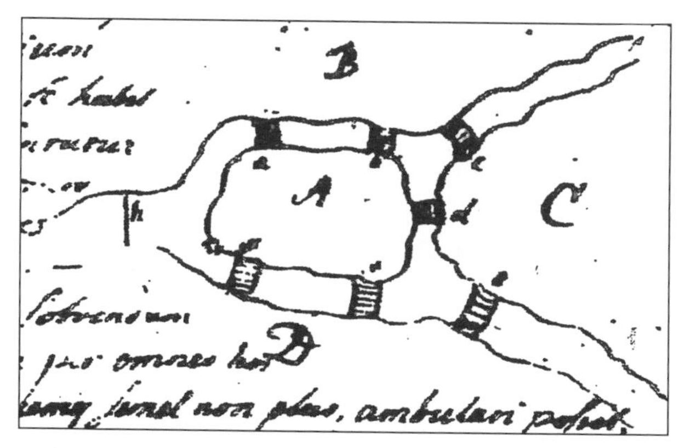

Euler's drawing of the seven bridges

LAGRANGE

Joseph-Louis Lagrange (1736–1813) excelled in all fields of analysis, number theory, and analytical and celestial mechanics. He wrote the first 'theory of functions', using the idea of a power series to make the calculus more rigorous, and his writings on mechanics were also highly influential. In number theory he proved that every positive integer can be written as the sum of at most four perfect squares.

Joseph-Louis Lagrange

Lagrange was born in Turin, Italy, to a family of Italian–French descent, and became professor of mathematics at the Royal Artillery School in Turin in 1755 at the age of only 19. At the invitation of Frederick the Great he succeeded Euler in Berlin in 1766, and remained there until Frederick's death in 1786. He spent the rest of his life in Paris.

Lagrange chaired the committee formed to introduce the metric system to France. He also took a leading role in the reform of university education, becoming professor at the École Normale in Paris in 1795, and at the École Polytechnique in 1797.

His early work contributed to the calculus of variations, which he applied to problems in dynamics. He also worked on the *libration* of the moon; this motion causes the face that the moon presents to the earth to oscillate slightly so that, over time, more than half of the moon's surface can be seen from the earth.

THE THEORY OF FUNCTIONS

Lagrange's two books on functions, *Théorie des Fonctions Analytiques* (1797) and *Leçons sur le Calcul des Fonctions* (1801), attempted to put calculus on to a more secure foundation by taking an algebraic approach rather than one based on a theory of limits. His method was to avoid any mention of tangents, or any use of pictures, and to define functions as infinite 'power series'. In particular, starting with a function that is written in the form

$$f(x) = a + bx + cx^2 + dx^3 + \ldots,$$

he *defined* its derivative to be

$$f'(x) = b + 2cx + 3dx^2 + \ldots;$$

for example, differentiating the function

$$\sin x = x - \tfrac{1}{6}x^3 + \tfrac{1}{120}x^5 - \ldots,$$

we get $1 - \tfrac{1}{3}x^2 + \tfrac{1}{24}x^4 - \ldots$, which is $\cos x$.

Although this approach to the foundations of the calculus turned out to be unsatisfactory, his abstract treatment of a function was a considerable advance. It essentially gave the first theory of functions of a real variable, with applications to a wide range of problems in geometry and algebra.

SOLVING POLYNOMIAL EQUATIONS

As we have seen, people have solved quadratic equations since Mesopotamian times, using only arithmetic operations (addition, subtraction, multiplication and division) and the taking of roots. In the 16th century, Italian mathematicians developed similar solutions to cubic equations (of degree 3) and quartic equations (of degree 4). We can solve all of these equations by means of formulas that involve only arithmetic operations and the taking of roots.

But how about equations of degree 5 (or more)? The corresponding search for a general solution or formula for these occupied the finest mathematicians, such as Descartes and Euler, but little progress was made on the problem until Lagrange attempted it, laying the groundwork for the eventual solution.

Lagrange's approach was to consider certain expressions involving the solutions of the equations (such as their sum or product) and to investigate how many different values these expressions can take when the solutions are permuted among themselves; for example, if the solutions of an equation are a, b and c, and the expression is $ab + c$, then we obtain *three* different values on permuting the solutions: $ab + c$, $ac + b$ and $bc + a$. Out of such explorations came a result that later became known, in a more general setting, as *Lagrange's theorem for groups.*

It was not until the 1820s that a proof finally emerged of the impossibility of solving the general equation of degree 5 (or more) by arithmetical operations and the taking of roots. This proof relied heavily on the ideas that Lagrange had put forward.

École Polytechnique students at the tomb of Gaspard Monge

MÉCANIQUE ANALYTIQUE

Lagrange's *Mécanique Analytique* (Analytical Mechanics) was his most important work. Published in 1788, just over a century after Newton's *Principia Mathematica,* it took a completely different approach to mechanics. It extended the work of Newton, the Bernoullis and Euler, and explained how one can generally answer questions about the motion of points and rigid bodies by reducing them to problems in the theory of ordinary and partial differential equations. As its contents page optimistically declared, it presents

> *Differential equations, for the solution of all the problems of Dynamics.*

In his *Mécanique Analytique* Lagrange transformed mechanics into a branch of mathematical analysis, and the geometrical approach that Newton had employed in the *Principia* was totally superseded. Indeed, as Lagrange emphasized in its preface:

> *One will not find figures in this work. The methods that I expound require neither constructions, nor geometrical or mechanical arguments, but only algebraic operations, subject to a regular and uniform course.*

LAPLACE

Pierre-Simon Laplace (1749–1827) was the last leading mathematician of the 18th century. He wrote a ground-breaking text on the analytical theory of probability, and is also remembered for the 'Laplace transform' of a function and Laplace's equation. His monumental five-volume work on celestial mechanics earned him the title of 'the Newton of France'.

Laplace was born in Normandy in France. Through the influence of d'Alembert he obtained a teaching position at the École Militaire in Paris, where legend has it that he examined (and passed) Napoleon. In 1790, during the French Revolution, he was appointed as a member of the committee of the Academy of Sciences formed to standardize weights and measures, and was subsequently involved with the organization of the École Normale and École Polytechnique.

Pierre-Simon Laplace

LAPLACE AND NAPOLEON

A well-known story, probably apocryphal, concerns the *Mécanique Céleste.* Summoned by Napoleon to give an account of his recently published book on the solar system, Laplace was asked by the Emperor why, unlike Isaac Newton, he had not mentioned God in his treatise. 'Sire', replied Laplace, 'I had no need of that hypothesis'.

THE *MÉCANIQUE CÉLESTE*

Laplace's *Traité de Mécanique Céleste* (Treatise on Celestial Mechanics), published in five volumes (the first two in 1799), consolidated the work of Newton, Clairaut, d'Alembert, Euler and Lagrange and his own researches. It was accompanied by an essay, *Exposition du Systeme du Monde* (Exposition on the System of the World) where he stated his philosophy of science:

> *If man were restricted to collecting facts the sciences were only a sterile nomenclature and he would never have known the great laws of nature. It is in comparing the phenomena with each other, in seeking to grasp their relationships, that he is led to discover these laws ...*

Laplace's work considered:

- the gravitational attraction of a spheroid on a particle outside it, and what we now know as *Laplace's equation* for the gravitational potential
- the motion of the moon
- the motion of three bodies under mutual gravitational attraction
- the perturbations of the planets and the stability of the solar system
- the nebular hypothesis for the formation of the solar system, arising from the contracting and cooling of a large mass of hot rotating gas.

DETERMINISM

Laplace believed in determinism, which he explains in the following quotation from the introduction to his non-technical *Essai Philosophique sur les Probabilités* (A Philosophical Essay on Probabilities):

> ***We may regard the present state of the universe as the effect of its past and the cause of its future. An intellect which at a certain moment would know all forces that set nature in motion, and all positions of all items of which nature is composed, if this intellect were also vast enough to submit these data to analysis, it would embrace in a single formula the movements of the greatest bodies of the universe and those of the tiniest atom; for such an intellect nothing would be uncertain and the future just like the past would be present before its eyes.***

282 ATTRACTIONS OF SPHEROIDS. [Méc. Cél.

therefore we shall have

[459]

$$0=\left(\frac{ddV}{dx^2}\right)+\left(\frac{ddV}{dy^2}\right)+\left(\frac{ddV}{dz^2}\right). \qquad (A)$$

Important Equation for computing the attractions of Spheroids and the figures of the Heavenly Bodies.

This remarkable equation will be of the greatest use to us, in the theory of the figures of the heavenly bodies. We may put it under other forms which are more convenient on several occasions.

[459']

Laplace's equation, from an English translation of *Mécanique Céleste*

PROBABILITY THEORY

Laplace's *Théorie Analytique des Probabilités* (Analytic Theory of Probability) of 1812 contains Laplace's definition of probability:

> *The theory of chance consists in the reduction of all events of the same kind to a certain number of equally likely cases that are cases such that we are equally undecided about their existence and in determining the number of cases which are favourable to the event whose probability is sought. The ratio to that of all the cases possible is the measure of this probability, which is thus simply the fraction whose numerator is the number of favourable cases, and whose denominator is the number of all cases possible.*

Laplace's statue in his birthplace of Beaumont-en-Auge

Laplace introduced generating functions for the solution of difference equations and also obtained approximations to binomial distributions. He worked on what is now called 'Bayes' theorem', which is relevant when an event can be produced by different causes: if the event happens, what is the probability that it was produced by a particular cause? As an example of his analysis he asked:

> *Over the period 1745–1770, 251,527 boys and 241,945 girls were born in Paris. Is this evidence that the probability of a male birth is greater than 0.5?*

His analysis showed, with extremely high probability, that this is indeed the case.

CHAPTER 4

THE AGE OF REVOLUTIONS

The 19th century saw the development of a mathematics profession in which people earned their living from teaching, examining and researching. The mathematical centre of gravity moved from France to Germany, while Latin gave way to national languages for publishing mathematical work. There was also a dramatic increase in the number of textbooks and journals.

Because of this increase in mathematical activity, mathematicians began to (indeed, needed to) specialize. While one would use the term *mathematician* in the 18th century, one now had *analysts, algebraists, geometers, number theorists, logicians* and *applied mathematicians.* This need for specialization was avoided only by the very greatest: Gauss, Hamilton, Riemann and Klein.

In each discipline there was a revolution (as well as an evolution) in the depth, extent, and even the very existence of the discipline. But each discipline experienced a movement towards an increasingly abstract style with an increased emphasis on putting mathematics on a sound and rigorous basis and examining its foundations. We illustrate this by considering the revolutions in three areas – analysis, algebra and geometry.

FROM CALCULUS TO ANALYSIS

In the 1820s Augustin-Louis Cauchy, the most prolific mathematician of the century, rigorized the calculus by basing it on the concept of a *limit.* He then used this idea to develop the areas of real and complex analysis. This increase in rigour necessitated the formulation of a foolproof definition of the real numbers, which in turn led to a study of infinite sets by Georg Cantor and others.

Joseph Fourier's work on heat conduction also gave rise to infinite processes — in this case, infinite series — thereby stimulating Bernhard Riemann in his work on integration. Analytical techniques came to be applied to a wide range of problems — in electricity and magnetism by William Thomson (Lord Kelvin) and James Clerk Maxwell, in hydrodynamics by George Gabriel Stokes, and in probability and number theory by Pafnuty Chebyshev.

The University of Göttingen, where Gauss, Riemann and Klein worked.

Revolutions did not happen only in mathematics: this is a miners' riot that took place in Belgium, 1868

FROM EQUATIONS TO STRUCTURES

Algebra also changed dramatically throughout the 19th century. In 1800 the subject was about solving equations, but by 1900 it had become the study of mathematical structures — sets of elements that are combined according to specified rules, called axioms.

At the beginning of the century, Gauss laid down the basics of number theory and introduced modular arithmetic, an early example of a new algebraic structure called a *group.*

A long-standing problem had to do with finding a general method for solving polynomial equations of degree 5 or more, using only arithmetical operations and the taking of roots. Niels Abel showed that there can be no such general solution, and Évariste Galois developed his ideas by examining groups of permutations of the roots of an equation.

The mystique concerning complex numbers was at last removed by William Rowan Hamilton, who defined them as pairs of real numbers with certain operations. Other algebraic structures were discovered: Hamilton introduced the algebra of quaternions, George Boole created an algebra for use in logic and probability, and Cayley studied the algebra of rectangular arrays of symbols, called *matrices*.

FROM ONE TO MANY GEOMETRIES

Over the space of one hundred years the study of geometry was completely transformed. In 1800 the only 'true' geometry had been Euclidean geometry, although there were some scattered results on spherical and projective geometry. By the end of the century, infinitely many geometries were known, while geometry had become closely linked to group theory and placed on a more rigorous foundation.

Gauss studied surfaces and their curvature, finding a relationship between curvature and the sum of the angles of a triangle on the surface, and this turned out to be related to the investigations into the parallel postulate in Euclidean geometry. Nikolai Lobachevsky and János Bolyai independently developed non-Euclidean geometry, in which the parallel postulate does not hold.

It took time, however, for the ideas on non-Euclidean geometry to become absorbed, and it was the mid-century work of Riemann that showed the importance of the new ideas and extended the work of Gauss. Through such abstract techniques, geometry was also moving out of two and three dimensions and into higher ones. Later, Felix Klein used groups to examine and classify different types of geometry.

GAUSS

Carl Friedrich Gauss (1777–1855) was one of the greatest mathematicians of all time. He made significant contributions to a wide variety of fields, including astronomy, geodesy, optics, statistics, differential geometry and magnetism. He presented the first satisfactory proof of the fundamental theorem of algebra and the first systematic study of the convergence of series. In number theory he introduced congruences and discovered when a regular polygon can be constructed with an unmarked ruler and pair of compasses. Although he claimed to have discovered a 'non-Euclidean geometry', he published nothing on it.

Gauss at his astronomical observatory in Göttingen

Gauss was born in the Duchy of Brunswick, now in Germany. A child prodigy, he reputedly summed all the integers from 1 to 100 by spotting that the total of 5050 arises from 50 pairs of numbers, with each pair summing to 101:

$$101 = 1 + 100 = 2 + 99 = \ldots = 50 + 51.$$

He went to Göttingen University in 1795, later returning to Brunswick, until he was appointed director of the Göttingen Observatory in 1807. He remained there for the rest of his life.

Gauss's *Disquisitiones Arithmeticae* (Discourses in Arithmetic) was published in 1801 when he was only 24. It was his most famous work, earning him the title of the 'Prince of Mathematics'. His view of number theory is captured in a famous quotation that is attributed to him:

Mathematics is the queen of the sciences, and number theory is the queen of mathematics.

CONSTRUCTING POLYGONS

In his teenage years Gauss became interested in the construction of regular polygons, using only an unmarked ruler and a pair of compasses. In the first proposition of Euclid's *Elements* we learn how to construct an equilateral triangle, and the *Elements* also contains constructions for a square and a regular pentagon.

We can also start with a regular n-sided polygon and construct a $2n$-sided polygon. For example, from an equilateral triangle we can construct regular polygons with 6, 12 and 24 sides, while a square gives polygons with 8, 16 and 32 sides, and a regular pentagon gives those with 10, 20 and 40 sides. But no-one can construct a regular polygon with 7 or 9 sides — so which regular polygons *are* constructible?

Gauss approached this question by first describing a complicated geometrical method for constructing a regular 17-sided polygon. He then analysed the general case, and came up with a

THE FUNDAMENTAL THEOREM OF ALGEBRA

The subject of Gauss's doctoral thesis, this theorem states that

Every polynomial factorizes completely into linear and quadratic factors.

It follows that every polynomial equation of degree n has n complex solutions.

surprising answer, involving the Fermat primes: the known ones are 3, 5, 17, 257 and 65537. He found that:

> *A regular n-sided polygon can be constructed if and only if n can be obtained by multiplying any number of different Fermat primes and then doubling as often as we wish.*

MODULAR ARITHMETIC

At the beginning of *Disquisitiones Arithmeticae* Gauss laid the foundations of number theory as a discipline with its own techniques and methods. To do so, he introduced modular arithmetic and congruences, a topic that exemplifies the rising abstraction of 19th-century mathematics.

For any positive integer n, two numbers a and b are said to be *congruent modulo n* if n divides $a - b$; n is called the *modulus* and we write $a \equiv b \bmod n$: so $37 \equiv 7 \bmod 10$, since 10 divides $37 - 7$. So if $n = 10$, we are dealing with the remainders 0, 1, 2, ... , 9 obtained after division by 10, because every integer is congruent mod 10 to just one of these.

Using his congruences, Gauss proved a famous result of Euler, known as the *quadratic reciprocity theorem:*

> *If p and q are odd prime numbers,*
> *then $x^2 \equiv p \bmod q$ has a solution*
> *if and only if $x^2 \equiv q \bmod p$ does,*
> *except when $p \equiv 3 \bmod 4$ and $q \equiv 3 \bmod 4$.*

ASTRONOMY AND STATISTICS

Also in 1801, the year of the *Disquisitiones Arithmeticae,* Gauss established himself as one of Europe's leading astronomers. On the first day of the century, Giuseppe Piazzi discovered the asteroid Ceres, the first new object discovered in the solar system since William Herschel had found Uranus twenty years earlier. Piazzi was able to observe it for only forty-two days before it disappeared behind the sun. But where would it reappear? Many astronomers gave their predictions, but only Gauss's was correct, thereby causing great excitement.

In his investigation of the orbit of Ceres, Gauss developed numerical and statistical techniques that would have lasting importance. In particular was his work on *the method of least squares,* which deals with the effect of errors of measurement. In this, he assumed that the errors in the measurements were distributed in a way that is now known as the *Gaussian* or *normal distribution.*

Ceres

GERMAIN

In the predominantly male world of late 18th-century university mathematics, it was difficult for talented women to become accepted. Discouraged from studying the subject, they were barred from admission to universities or the membership of academies. One mathematician who had to struggle against such prejudices was Sophie Germain (1776–1831).

Germain was born in Paris, the daughter of a wealthy merchant who later became a director of the Bank of France. Her interest in mathematics supposedly began during the early years of the French Revolution.

Confined to her home because of rioting in the city, she spent much time in her father's library. Here she read an account of the death of Archimedes at the hands of a Roman soldier, and determined to study the subject that had so engrossed him. But her parents were strongly opposed to such activities, believing them to be harmful for young women. At nighttime they even removed her heat and light and hid her clothes to dissuade her, but she persisted and they eventually relented.

Sophie Germain, aged 14

MONSIEUR LE BLANC

During the Reign of Terror in France, Sophie Germain remained at home, teaching herself the differential calculus. In 1794, when she was 18, the École Polytechnique was founded in order to train much-needed mathematicians and scientists. This would have been the ideal place for her to study, but it was not open to women.

Frustrated, but undeterred, she decided on a plan of covert study. She managed to obtain the lecture notes for Lagrange's exciting new course on analysis, and at the end of the term submitted a paper under the pseudonym of M. Antoine Le Blanc, a former student of the École.

Lagrange was so impressed by the originality of this paper that he insisted on meeting its author. When Germain nervously turned up, he was amazed, but delighted. He proceeded to give her much help and encouragement, putting her in touch with other French mathematicians and helping her to develop her mathematical interests.

One of the most important of these was the theory of numbers. Germain wrote to Adrien-Marie Legendre, the author of a celebrated book on the subject, about some difficulties she had found with his book. This led to a lengthy and fruitful exchange.

Another productive correspondence was with the great Gauss. His recent *Disquisitiones Arithmeticae* on number theory had impressed Sophie Germain so much that she plucked up the courage to send him her discoveries, once again choosing to present herself as Monsieur Le Blanc of the École Polytechnique.

FERMAT'S LAST THEOREM

One topic that Sophie Germain included in her communications with Gauss was Fermat's last theorem — that, for any integer n (> 2), there do not exist any positive integers x, y and z for which $x^n + y^n = z^n$. Fermat had claimed to have a general proof of this, and presented one for the case $n = 4$, and Euler subsequently proved it for the case $n = 3$, but no other results were known at the time.

Over the next few years, Germain obtained several new results on Fermat's last theorem, proving in particular that, if n is any prime number less than 100, then there are no positive integer solutions if x, y and z are mutually prime to one another and to n.

Some Chladni patterns

Gauss was impressed with Germain's discoveries and continued to correspond with her. Her identity remained secret until 1807, when French troops occupied the city of Hanover, where Gauss resided. Fearful that Gauss might suffer a fate similar to that of Archimedes, she contacted the French commander, General Pernety, a family friend, who agreed to guarantee Gauss's safety and revealed to him the source of the request. Gauss wrote to her of his surprise and delight, praising her for her 'noblest courage, quite extraordinary talents and a superior genius'.

ELASTICITY

When Gauss went to Göttingen in 1808, Sophie Germain lost interest in number theory and, inspired by some lectures by the German physicist Ernst Chladni, she became involved with elasticity and acoustics; Chladni had scattered sand on a glass plate and observed the patterns that appear when a violin bow is drawn along the edge of the plate.

Such observations had no known theoretical basis, and the French Academy of Sciences offered a prize for formulating a mathematical theory for elastic surfaces and explaining how it agreed with observation. Using some of Sophie Germain's results, Lagrange discovered the partial differential equation for the vibrations of a flat plate, from which she developed a general theory of vibrations of a curved surface. This so impressed the judges that she was awarded the prestigious prize and a medal from the Institute of France. Her work in this area subsequently provided the basis for the modern theory of elasticity.

Arising from her work on curved surfaces, Germain was led to study their curvatures — these are numbers that describe how 'curved' the surface is in different directions. Her last major achievement was to define the *mean curvature* of a surface, a concept that has since been of importance in the geometry of surfaces.

MONGE AND PONCELET

Out of the turbulent years of the French Revolution and the rise to power of Napoleon Bonaparte came important mathematical developments. One of Napoleon's greatest supporters was the geometer Gaspard Monge (1746–1818), who advised him on fortifications on his Egyptian expedition of 1798. Monge's most talented student was Jean Victor Poncelet (1788–1867), the 'Father of Modern Projective Geometry'.

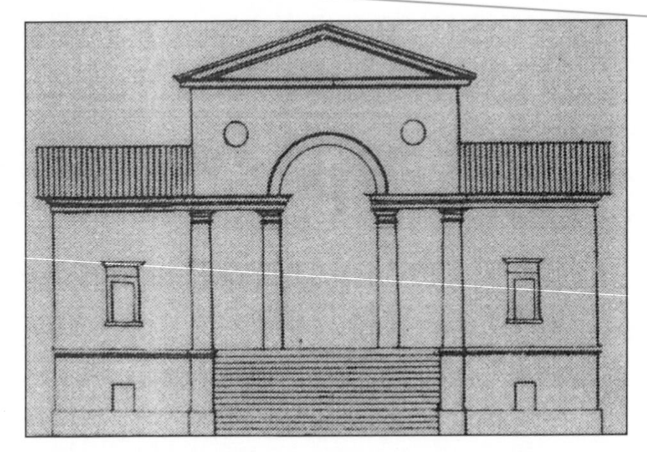

A drawing from J. Durand's book on descriptive geometry (1802–1805)

Napoleon was an enthusiast for mathematics and its teaching, and there is a result in geometry that is commonly attributed to him:

> *Given any triangle ABC, construct equilateral triangles on its sides and join the centres of gravity of these three triangles — then the resulting triangle is always equilateral.*

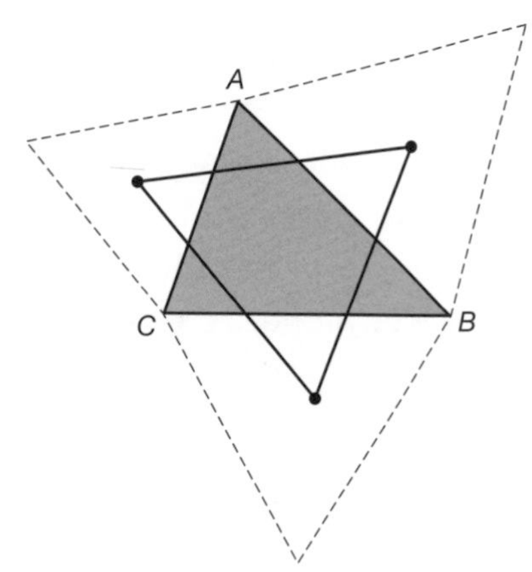

Napoleon's theorem

An important consequence of the French Revolution was Napoleon's founding of the École Normale and the École Polytechnique in Paris. In these institutions, the country's finest mathematicians, including Monge, Lagrange, Laplace and Cauchy, taught the students who were destined to serve in both military and civilian capacities. The textbooks produced by the École teachers were later to be widely used in France and the United States.

MONGE

Napoleon's close friend, the geometer Gaspard Monge, taught at the military school in Mézières, where he studied the properties of lines and planes in three-dimensional Cartesian geometry. While investigating possible positionings for gun emplacements in a fortress, he greatly improved on the known methods for projecting three-dimensional objects on to a plane; this subject soon became known as 'descriptive geometry'.

Monge was also involved with 'differential geometry', in which techniques of calculus are used to study curves drawn on surfaces, and wrote the first important book on the subject.

Gaspard Monge's skill as a teacher at the École Polytechnique, where he was director, helped to establish descriptive geometry and to inspire his talented students. Although his approach was mainly practical, he also developed the necessary algebraic machinery to render the subject rewarding and versatile. After Napoleon's exile in 1815, he lost his Polytechnique position and died soon after.

PROJECTIVE GEOMETRY

We have seen how the properties of perspective were investigated by Renaissance painters, and later by Desargues and Pascal, giving rise to some interesting results.

In Euclid's geometry,

Any two points determine a unique line,
and any two lines meet at a unique point
(*unless they are parallel*).

This exception about parallel lines seems awkward, and we are tempted to see what happens if we amend these statements to:

Any two points determine a unique line,
and any two lines meet at a unique point.

We can then think of parallel lines meeting at a 'point at infinity', although this point is not to be considered as any different from any other.

Parallel lines meeting at infinity

This produces a completely different kind of geometry, known as *projective geometry.* In particular, as noticed by Poncelet and his Paris contemporary Joseph Gergonne, there is now a *duality* between points and lines: any result concerning points lying on lines can be 'dualized' into another one about lines passing through points, and conversely — prove one and get another for free!

This revolutionary idea provided a real break with the past. As profound as it was controversial, it caused difficulties that the French geometers were unable to resolve.

PONCELET

Following Napoleon's ill-fated invasion of Russia in 1812, Monge's student Jean Victor Poncelet was thrown into a Russian prison. While languishing there, he developed the idea of a 'projective transformation' (such as projecting a diagram on to a screen from a point light source), and investigated those geometrical properties of figures that remain unchanged by such transformations. His work was greatly influenced by that of Desargues and Monge, but was intuitive and unrigorous and was poorly received by the mathematical hierarchy in Paris.

Poncelet's treatise on the projective properties of figures appeared in 1822:

This book is the result of researches which I undertook in the spring of 1813 in the prisons of Russia: deprived of every kind of books and help, and the proper facilities, and above all distracted by the misfortunes of my country, I was unable to give it all the perfection desirable. However, I had at the time found the fundamental theorems in my work: that is to say, the principles of central projection of figures ...

Jean Victor Poncelet

CAUCHY

As we have seen, the foundations of the calculus were found to be shaky and d'Alembert and others attempted to rescue them. The difficulties were overcome by Augustin-Louis Cauchy (1789–1857), who was the leading mathematician in France and the most important analyst of the early 19th century. In the 1820s he transformed the subject by formalizing the concepts of limit, continuity, derivative and integral. In addition, he helped to develop the algebraic idea of a 'group' and almost single-handedly created the subject of complex analysis.

After training as a civil engineer, Cauchy went to Cherbourg where he worked on designs for the harbour and the fortifications. His first mathematical papers were on polyhedra and algebra. He was soon elected to the Academy of Sciences and moved to Paris, where he lectured at the École Polytechnique.

BOLZANO

In 1817, in his home city of Prague, a Catholic priest named Bernard Bolzano produced a pamphlet, snappily entitled *A Purely Analytic Proof of the Theorem that, Between any Two Values that give Results of Opposite Sign Lies at least One Real Root of the Equation.* This result is now called the *Intermediate value theorem,* and tells us that:

> *A continuous graph that is below the x-axis at one place and above it at another must cross the x-axis at some point in between.*

Although this result may seem obvious, Bolzano's pamphlet contained its first rigorous proof.

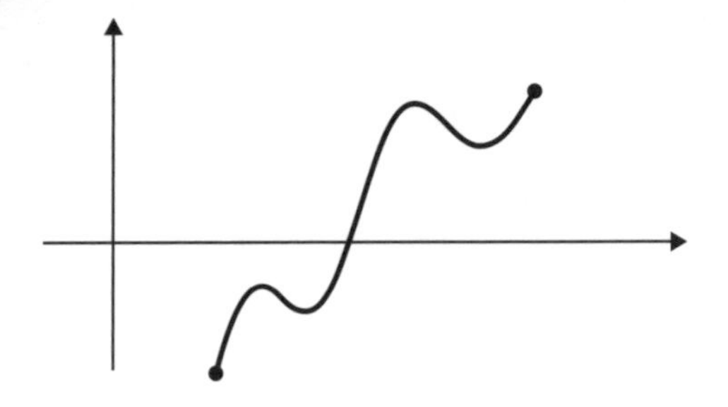

COURS D'ANALYSE

DE

L'ÉCOLE ROYALE POLYTECHNIQUE;

PAR M. AUGUSTIN-LOUIS CAUCHY,

Ingénieur des Ponts et Chaussées, Professeur d'Analyse à l'École polytechnique, Membre de l'Académie des sciences, Chevalier de la Légion d'honneur.

I.re PARTIE. *ANALYSE ALGÉBRIQUE.*

DE L'IMPRIMERIE ROYALE.

Chez DEBURE frères, Libraires du Roi et de la Bibliothèque du Roi, rue Serpente, n.° 7.

1821.

Intuitively, a graph is 'continuous' if it has no gaps, but Bolzano found it necessary to formalize this idea:

> *A function $f(x)$ varies continuously for all values of x in a certain interval if, for any x in that interval, the difference $f(x + \omega) - f(x)$ can be made smaller than any given quantity by insisting that ω be taken as small as we please.*

For example, if $f(x) = x^2$ between 0 to 1, then, for any value of x in this range,

CAUCHY'S COMPLEX ANALYSIS

We have seen how Leibniz defined the integral $\int_a^b f(x)\,dx$ of a function *f* as a 'sum of lines' – we can think of this informally as the result of 'adding up all the values of *f*(*x*)' as *x* travels from *a* to *b*.

In the late 1820s Cauchy explained how this idea can be extended to complex numbers. If *f*(*z*) is a function of a complex variable *z* (such as $f(z) = z^2$), and if *P* is a curve in the complex plane, then we can analogously define $\int_P f(z)\,dz$ as the result of 'adding up all the values of *f*(*z*)' as *z* travels along the curve *P*.

Cauchy proved many spectacular results concerning these complex integrals. The most powerful of these results involves integrating functions that are differentiable around closed curves (curves whose ends coincide, such as those shown).

Known as *Cauchy's theorem*, it tells us that $\int_P f(z)\,dz = 0$ whenever *f* is differentiable and *P* is a closed curve.

Some closed paths

Furthermore, *Cauchy's integral formula* states that if *a* is any point inside the closed curve *P*, then

$$f(a) = 1/(2\pi i) \int_P f(z) \,/\, (z - a)\,dz$$

— this tells us that we can find the value of *f*(*a*) at any point *a inside P* from the values of *f*(*z*) at all points *z on P*. It is rather like calculating the temperature at an inland place like Birmingham when we are told the temperatures at all points along the British coastline — a remarkable result.

$$f(x + \omega) - f(x) = (x + \omega)^2 - x^2 = \omega\,(2x + \omega),$$

which can be made as small as we please by choosing ω to be small enough. So $f(x) = x^2$ is continuous on this interval.

However, Bolzano's work never had the credit it deserved, as Prague was situated far from the centres of mathematical activity.

CAUCHY'S *COURS D'ANALYSE*

Meanwhile, much progress was being made in Paris. In 1821, Cauchy produced a ground-breaking book entitled *Cours d'Analyse* (Course of Analysis) in which he formalized the idea of a limit:

> *When the values successively attributed to the same variable approach a fixed value indefinitely, in such a way as to end up by differing from it as little as we could wish, this last value is called the limit of all the others.*

For example, $f(x) = (\sin x)/x$ is not defined when $x = 0$ (since 0/0 is meaningless), but as x approaches the value 0, $f(x)$ approaches the value 1, so 1 is the limit in this case.

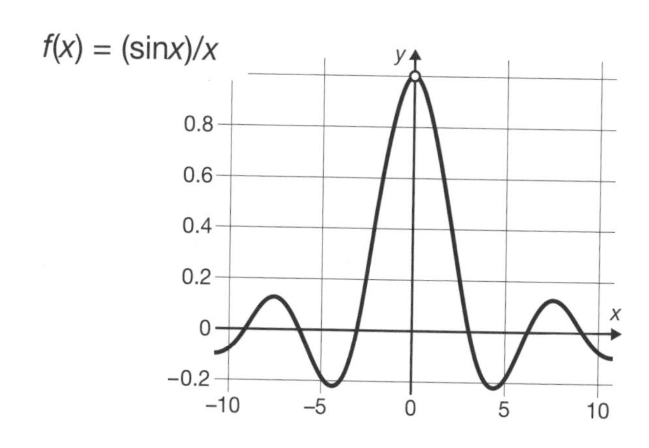

Using this definition, Cauchy was able to transform the whole subject. He gave rigorous explanations of what it means to say that a graph is *continuous* (no gaps) and *smooth* (no corners), and also gave rigorous presentations of the two fundamental concepts of the calculus, *differentiation* and *integration*.

FOURIER AND POISSON

Joseph Fourier

Joseph Fourier (1768–1830) worked on what are now known as *Fourier series*: this led to many of the most important mathematical discoveries of the 19th century, and had major applications in mathematical physics. Siméon Denis Poisson (1781–1840) has his name attached to *Poisson's equation* in potential theory and the *Poisson distribution* in probability theory.

Fourier was born in Auxerre in Burgundy, and in 1797 succeeded Lagrange in the chair of analysis and mechanics at the École Polytechnique, leaving the following year with Monge to join Napoleon's invasion of Egypt as a scientific adviser. On his return he was appointed by Napoleon to an administrative position at Grenoble in south-eastern France, organizing the draining of the swamps of Bourgoin and supervising the building of the road from Grenoble to Turin. In his spare time he carried out his important mathematical research on the conduction of heat.

THE CONDUCTION OF HEAT

In his 1822 *Théorie Analytique de la Chaleur* (Analytic Theory of Heat), Fourier wrote:

> *Fundamental causes are not known to us; but they are subject to simple and constant laws, which one can discover by observation and whose study is the object of natural philosophy.*

Fourier began his investigations into heat by obtaining a partial differential equation for the equilibrium temperature distribution in a rectangular region, where the temperatures at the boundaries are kept constant. This led to his deriving a representation of a *square wave* in terms, not of a power series, but of the infinite trigonometric series

$\cos u - \frac{1}{3}\cos 3u + \frac{1}{5}\cos 5u - \frac{1}{7}\cos 7u + \ldots,$

which equals 0 when $u = \pi/2$,

$\pi/4$ when u lies between $-\pi/2$ and $\pi/2$,

and $-\pi/4$ when u lies between $\pi/2$ and $3\pi/2$.

He wrote about this surprising outcome:

> *As these results appear to depart from the ordinary consequences of the calculus, it is necessary to examine them with care, and to interpret them in their true sense.*

Fourier then considered the more general question of which functions can be represented by Fourier series, first defining what he meant by a function:

> *In general, a function f(x) represents a succession of values or ordinates each of which is arbitrary. An infinity of values being given to the abscissa x, there is an equal number of ordinates f(x). ... We do not suppose these ordinates to be subject to a common law: they succeed each other in any manner whatsoever, and each of them is given as if it were a single quantity.*

Fourier did not consider functions as general as those described in his definition, but rather as those that are given by different rules over

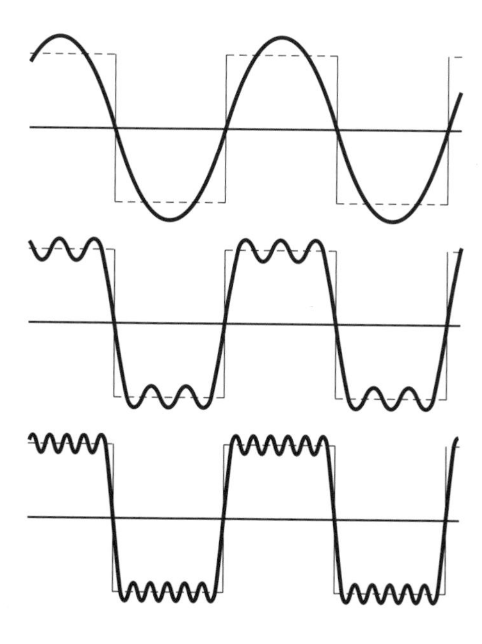

Approximating a square wave by a Fourier series

different sections where they are defined. He also derived formulas (involving integrals) for the coefficients in the Fourier series of the function.

The question of what conditions should be imposed on a function so as to ensure that its Fourier series does indeed converge to the original function generated much new activity, by Abel and Riemann amongst others.

POISSON

Poisson was born in Pithiviers in north-central France and quickly achieved academic success, occupying many educational positions — in particular, succeeding Fourier in 1806 at the École Polytechnique after Napoleon had sent Fourier to Grenoble. He published many mathematical works and, according to François Arago, frequently said:

> *Life is good for only two things, discovering mathematics and teaching mathematics.*

Poisson carried out major work on electricity, magnetism and elasticity, obtaining (for example) a partial differential equation giving the electric potential for a given distribution of the electric charges. In 1812 he won the French Academy's Grand Prix, where the topic was:

> *To determine by calculation and to confirm by experiment the manner in which electricity is distributed at the surface of electrical bodies considered either in isolation or in the presence of each other — for example, at the surface of two electrified spheres in the presence of each other.*

In his 1838 *Recherches sur la Probabilité des Jugements en Matière Criminelle et en Matière Civile* (Researches on the Probability of Judgments in Criminal and Civil Matters), he introduced what is now known as the *Poisson distribution.* This important distribution gives the probabilities of how many times an event occurs in an interval of time or a region of space. It makes certain assumptions that the events occur independently, and about how many occur in a short interval of time or a small region of space. He also introduced the term *law of large numbers.*

Siméon Denis Poisson

ABEL AND GALOIS

The tragic stories of Niels Henrik Abel (1802–1829) and Évariste Galois (1811–1832) are depressingly similar. Both found it difficult to get their results accepted, and although both made major advances in the theory of equations – Abel proved that no general solution can exist for polynomial equations of degree 5 or more, while Galois determined when such equations *can* be solved – both died young, Abel from tuberculosis and Galois after being wounded in a duel.

Earlier we saw that polynomial equations of degrees 2, 3 and 4 had been solved with only arithmetic operations and the taking of roots, but that no-one had been able to do the same for general equations of higher degrees. We also saw Lagrange's new approach to such problems, where he counted the number of different expressions that one can obtain by permuting the solutions of the given equation.

ABEL

Growing up in Norway, Abel was desperate to study in the main centres of mathematical life in France and Germany, and was eventually able to obtain a stipend that enabled him to spend time in Paris and Berlin.

In Germany he met Leopold Crelle and published many papers in the early issues of Crelle's new journal, thereby helping it to become the leading German mathematical periodical of the 19th century; among these papers was the one that contained his proof of the impossibility for solving the general equation of degree 5 or more. He also obtained fundamental results on other topics (the convergence of series, elliptic functions, and 'Abelian integrals'), many of which appeared in his 'Paris memoir' of 1826.

Niels Henrik Abel

The story of Abel's attempts to be recognized by the mathematical community, and of his lack of success in securing an academic post, is a sorry one. For a time, his Paris memoir was lost. He then returned to Norway where he contracted tuberculosis and died at the early age of 26. Two days later, a letter arrived at his home, informing him that his memoir had been found and offering him a prestigious professorship in Berlin.

GALOIS

The work of Lagrange and Abel on the unsolvability of the general quintic equation was developed by the brilliant Évariste Galois, who determined criteria (in terms of an object now called the *Galois group*) for deciding *which* polynomial equations can be solved by arithmetical operations and the taking of roots. His work ultimately led to whole new areas of algebra, now known as *group theory* and *Galois theory.*

Galois's teenage years were traumatic. He failed his entrance examination for the École Polytechnique. A manuscript that he sent to the French Academy of Sciences was mislaid, another was rejected for being obscure, and his father committed suicide.

Évariste Galois

A republican firebrand who became involved with political activities following the July Revolution of 1830, Galois threatened the life of King Louis-Philippe, but was acquitted. A month later he was discovered carrying weapons and wearing the uniform of the banned artillery guard, whereupon he was thrown into jail.

Galois spent the night before his duel frantically scribbling a letter to his friend August Chevalier, summarizing his results and requesting Chevalier to show them to Gauss and Jacobi. But it was to be several years before anyone appreciated what they meant, and what a genius the world had lost.

APPLYING 19TH-CENTURY ALGEBRA

As we have seen, the Greeks were fascinated by geometrical constructions. Using only an unmarked ruler and a pair of compasses, they bisected angles, trisected line segments, and constructed squares with the same area as a given polygon. But there were three types of construction that defeated them:

Doubling a cube

Given a cube, construct another cube with twice the volume.

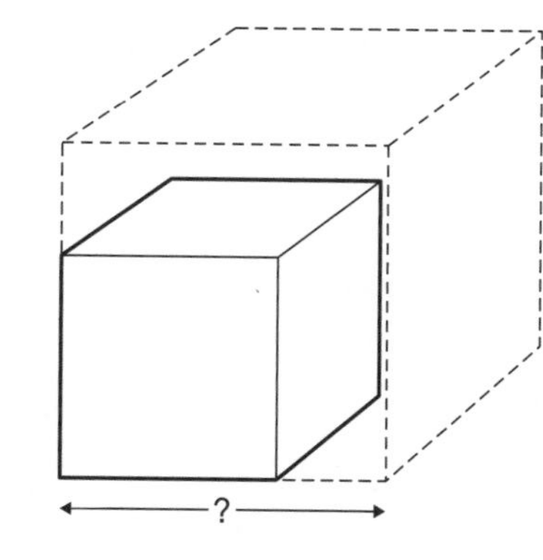

Trisecting an angle

Given any angle, divide it into three equal parts.

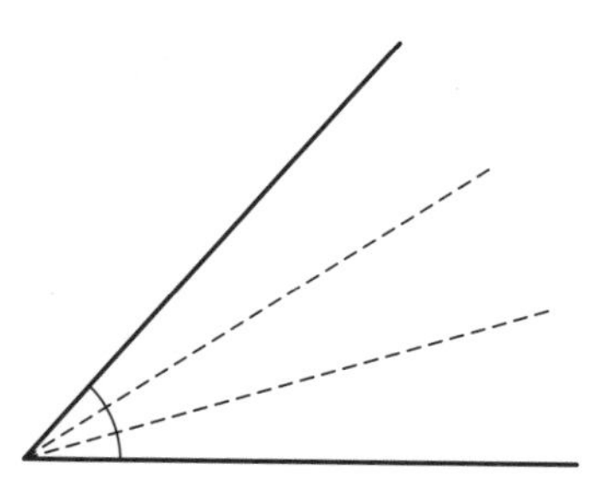

Squaring the circle

Given any circle, construct a square with the same area.

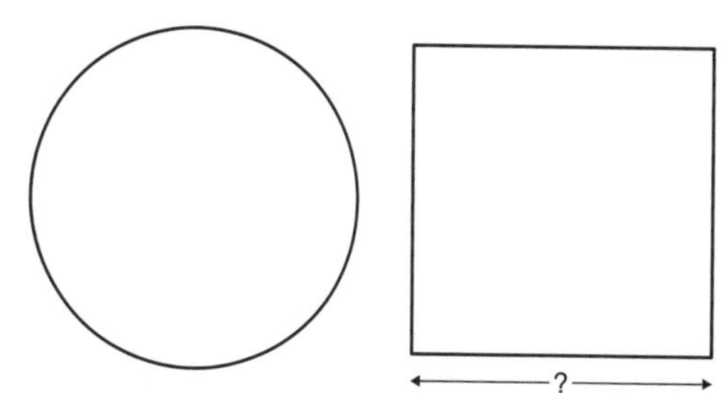

These all date from the 4th century BC, and for the next two millennia valid constructions were sought, without success.

Starting with a line of length 1, we can mark out its multiples, and then construct lines of any lengths that are rational numbers (fractions). By intersecting these lines with circles, and intersecting circles with other circles, we find that:

We can construct any lengths involving basic arithmetic operations and the successive taking of square roots, but no other lengths.

DOUBLING A CUBE

If the first cube has side 1, then the doubled cube has side $\sqrt[3]{2}$, which is a cube root and so cannot be constructed.

TRISECTING AN ANGLE

If we try to trisect an angle of 60°, we find that $x = \cos 20°$ satisfies the equation $8x^3 - 6x - 1 = 0$, whose solutions involve cube roots that cannot be constructed.

SQUARING THE CIRCLE

This involves π, which also cannot be constructed.

Thus, once 19th-century algebraists had proved that none of the lengths $\sqrt[3]{2}$, cos 20° and π can be constructed, it followed that:

All three types of construction are impossible.

MÖBIUS

The 1820s and 1830s witnessed a shift in mathematical activity from France to Germany, with the Paris Écoles giving way to the Universities of Berlin and Göttingen. But several mathematicians, such as Gauss, were also involved with astronomy and were employed in observatories, rather than universities. August Möbius (1790–1868) combined the posts of Professor of Astronomy at Leipzig University and Director of the Observatory, while carrying out his wide-ranging mathematical pursuits.

Möbius was born in Schulpforta in Saxony and studied at Leipzig University, before travelling to Göttingen to study astronomy with Gauss. His doctoral thesis was on the occultations of fixed stars, and his Leipzig 'habilitation' (enabling him to teach in the University) was on trigonometrical equations. He was appointed Professor of Astronomy at the University of Leipzig and the

Leipzig Observatory

Observatory was developed under his supervision. In addition to his University teaching post, he became Observer at the Observatory in 1816, being promoted to Director in 1848.

BARYCENTRIC COORDINATES

Although modern projective geometry was born in France, arising out of the work of Poncelet, the scene soon moved to Germany with contributions by Möbius and others. In 1827, Möbius introduced algebraic methods into projective geometry, just as Descartes and his successors had done for analytic geometry two centuries earlier when they represented points by number pairs (a, b) and lines by equations of the form $ax + by + c = 0$.

To do this, Möbius introduced *barycentric coordinates.* Consider an object attached to three strings that pass through holes A, B and C in a table. If weights a, b and

August Möbius

THE FIVE PRINCES

In his classes at Leipzig around 1840, Möbius asked the following question of his students:

There was once a king with five sons. In his will he stated that after his death the sons should divide the kingdom into five regions in such a way that each one should share part of its boundary with each of the other four regions. Can the terms of the will be satisfied?

This is one of the earliest problems from the area of mathematics that is now known as *topology.*

The answer to the question is *no.*

MÖBIUS TRANSFORMATIONS

There are many ways of transforming the complex plane into itself. For example:

- **the transformation $f(z) = (1 + i)z$ has the effect of rotating and expanding square grids of lines:**

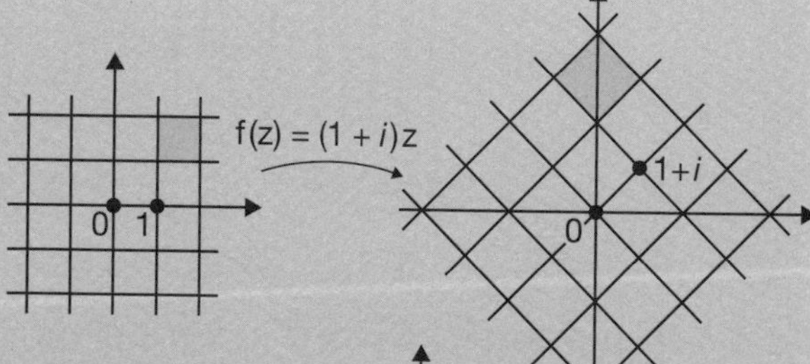

- **the transformation $f(z) = 1/z$ transforms horizontal and vertical lines into circles:**

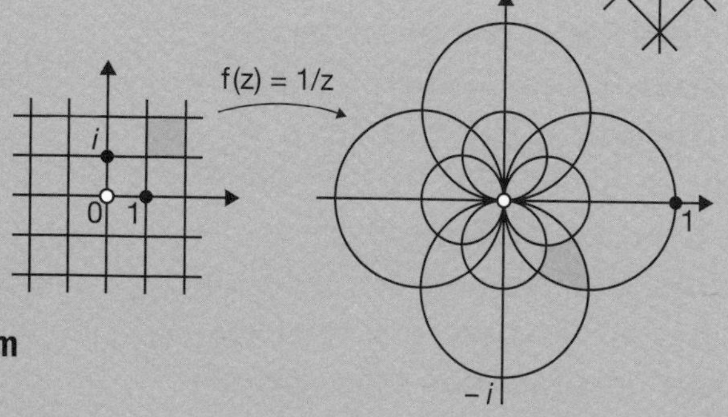

These are special cases of what we call *Möbius transformations*, which have the form

$f(z) = (az + b) / (cz + d)$, where $ad \neq bc$.

These very versatile transformations enable us to transform chosen areas of the plane to other areas; for example, we can transform the right-hand half of the plane to the interior of the circle with radius 1 by means of the transformation

$f(z) = (z - 1) / (z + 1)$:

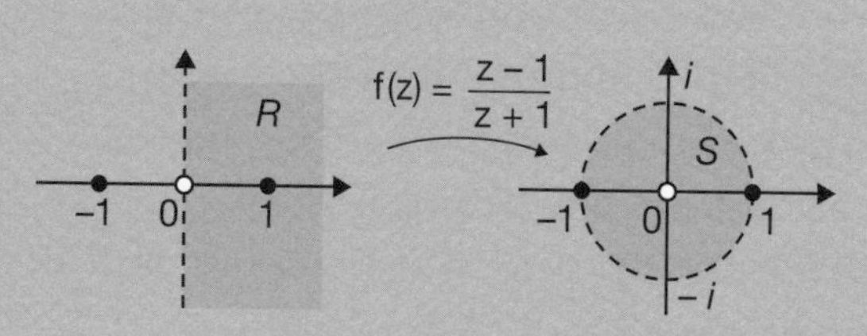

c are attached to the strings, then the object finds equilibrium at a point P inside the triangle ABC, to which we assign the coordinates $[a, b, c]$.

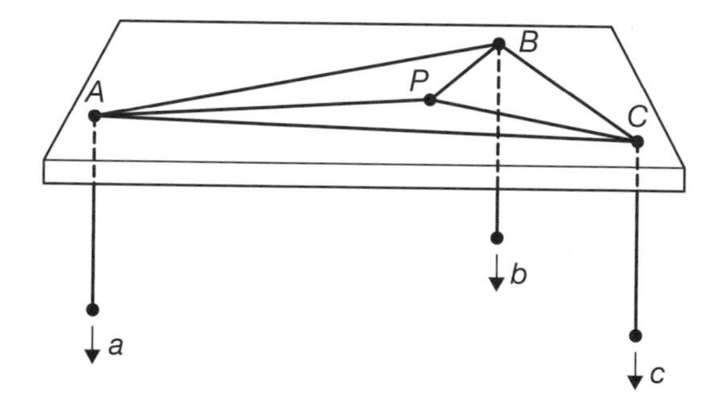

If we now double all the weights, or multiply them by any other fixed number, then the point P stays fixed. But in general, different triples of weights give rise to different points; also, no point corresponds to the triple of weights $[0, 0, 0]$. Möbius then showed how to obtain points outside the triangle ABC by allowing weights to take negative values.

Thus we obtain a geometry in which the points are triples of numbers $[a, b, c]$ (other than $[0, 0, 0]$), defined up to multiples. We can also define lines in this geometry as equations of the form $ax + by + cz = 0$. Points and lines are then related by the duality

$[a, b, c] \leftrightarrow ax + by + cz = 0$.

THE MÖBIUS BAND

Möbius is probably best remembered for describing the object known as a *Möbius band* (or *Möbius strip*) in 1858. To construct one, take a long strip of paper, twist one end through 180°, and then glue its ends together.

A Möbius band has several unexpected properties. For example, it has only one side, as you will discover if you trace along the middle of it with a pencil until you reach your starting point. If you then cut it along this line, you obtain two linked paper rings.

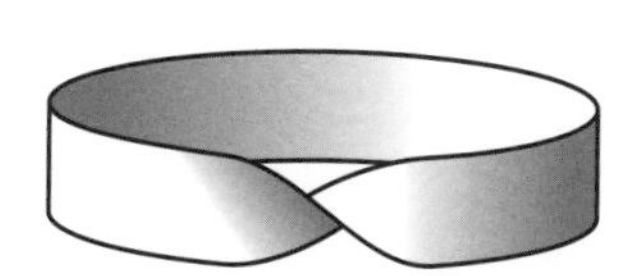

BOLYAI AND LOBACHEVSKY

As we have seen, Euclid's *Elements* is built on five self-evident truths, called *postulates*. Four of these are straightforward, but the fifth is different in style. For 2000 years people tried to deduce it from the other four, but no-one could do so. This is because there are 'non-Euclidean' geometries that satisfy the first four postulates, but not the fifth. Their existence was first published around 1830 by the Transylvanian János Bolyai (1802–1860) and the Russian Nikolai Lobachevsky (1792–1856).

Over the centuries, many tried to prove the fifth postulate by deducing it from other results in Euclid's *Elements* to which it is equivalent. Two of these were the 'parallel postulate':

Given any line L and any point P that does not lie on this line, there is exactly one line, parallel to L, that passes through P

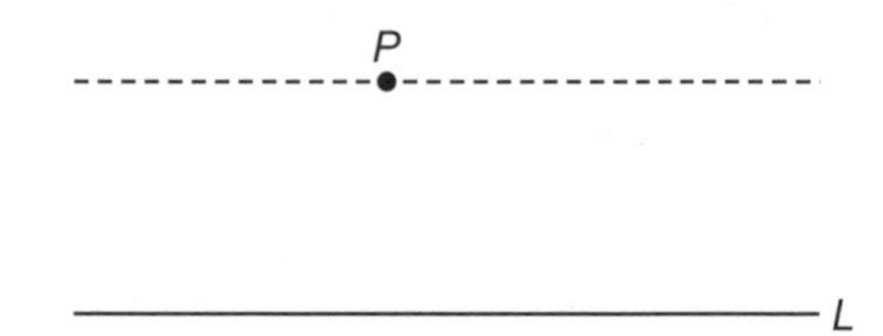

and the 'angle-sum theorem for rectangles':

The angles of any rectangle add up to 360°.

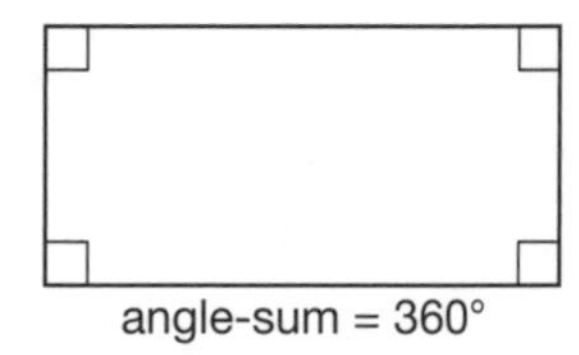
angle-sum = 360°

If one can deduce either of these from the first four postulates, then the fifth one must also be true; earlier we described Alhazen's unsuccessful attempt to prove the parallel postulate.

SACCHERI'S ATTEMPT

The first really significant advance was in 1733, in *Euclides ab Omni Naevo Vindicatus* (Euclid Vindicated from Every Flaw), by the Italian geometer Gerolamo Saccheri. His approach was to consider geometries in which the fifth postulate is not assumed, and derive a contradiction.

In order to do so, Saccheri attempted to prove that rectangles whose angle-sum is not equal to 360° cannot exist. It would then follow that the angle-sum is always 360° and that the fifth postulate is true.

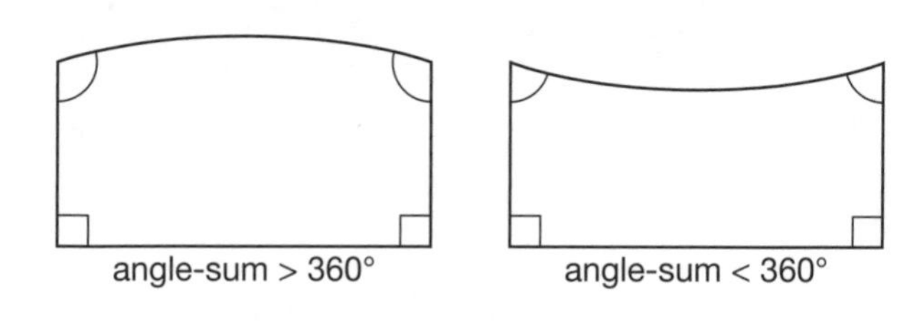
angle-sum > 360° angle-sum < 360°

Saccheri's first attempts were successful. He proved that if the angle-sum is greater than 360°, then the parallel postulate can be proved both true and false — as Saccheri remarked:

It is absolutely false, because it destroys itself.

This contradiction shows that no geometry can have this property.

He then tried to repeat the process for a rectangle with angle-sum less than 360°, claiming that:

The hypothesis of the acute angle is absolutely false, because it is repugnant to the nature of straight lines.

But here his argument contained an error.

If Saccheri *had* been successful in this case, then he would have proved that the sum of the angles in every rectangle must be 360°. It would then follow that the fifth postulate can indeed be deduced from the other postulates.

NON-EUCLIDEAN GEOMETRIES

Saccheri's approach was proved wrong in spectacular fashion. Around 1830, Bolyai and Lobachevsky independently both constructed a new type of geometry in which

> *The angles of any rectangle add up to less than* 360°.

In their geometry the first four of Euclid's postulates still hold, but not the fifth one.

This Bolyai–Lobachevsky geometry has some very strange features.

> *Given any line L and any point P that does not lie on this line, there are infinitely many lines, parallel to L, that pass through P.*

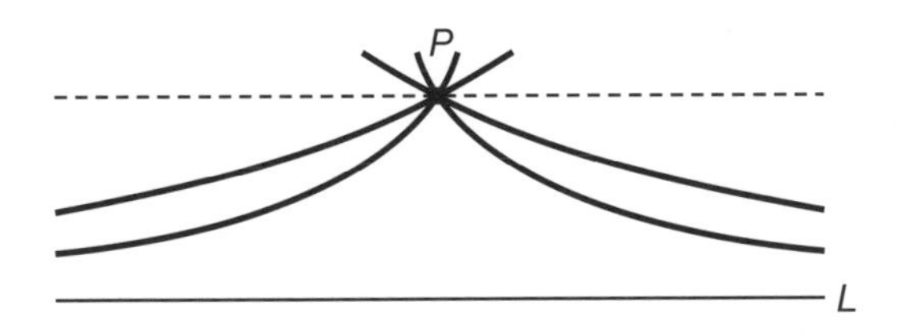

Moreover, *if two triangles are similar* (they have the same angles), *then they must also be congruent* (they have the same size) — which is also not true in Euclidean geometry.

Both Bolyai and Lobachevsky were unsuccessful in getting their work widely known, and they enjoyed little of the credit due to them for their remarkable discovery. It was not until after their deaths that their geometries were fully understood.

THE CONTRIBUTION OF GAUSS

The non-Euclidean geometries of Bolyai and Lobachevsky were considered very controversial, as they were seen not to correspond to the world we live in. Some years earlier, Gauss had been thinking along similar lines:

> *I am becoming more and more convinced that the necessity of our [Euclidean] geometry cannot be proved ... Perhaps in another life we will be able to obtain insight into the nature of space, which is now unattainable.*

However, he was unwilling to publish his startling predictions, fearing 'the howl of the Boeotians' if he did so; the Boeotians were ancient Greeks who were resistant to change.

János's father, Farkas Bolyai, had also worked on the parallel postulate and earnestly tried to dissuade his son from doing so:

> *You must not attempt this approach to parallels. I know its way to its very end. I have traversed this bottomless night, which extinguished all light and joy of my life ... I have travelled past all reefs of this infernal Dead Sea and have always come back with broken mast and torn sail.*

But the son persisted, and when Farkas Bolyai informed his old friend Gauss of his son's success, Gauss accepted the results but claimed them as his own:

> *If I commenced by saying that I am unable to praise this work, you would certainly be surprised for a moment. But I cannot say otherwise. To praise it would be to praise myself. Indeed the whole contents of the work, the path taken by your son, the results to which he is led, coincide almost entirely with my meditations, which have occupied my mind partly for the last thirty or thirty-five years.*

Nikolai Lobachevsky

János Bolyai never forgave Gauss for this.

BABBAGE AND LOVELACE

The central figure of 19th-century computing was Charles Babbage (1791–1871), who may be said to have pioneered the modern computer age with his 'difference engines' and his 'analytical engine', although his influence on subsequent generations is hard to assess. Ada, Countess of Lovelace (1816–1852), daughter of Lord Byron and a close friend of Babbage, produced a perceptive and clear commentary on the powers and potential of the analytical engine; this was essentially an introduction to what we now call programming.

A portion of the 1832 difference engine: it was to have the feature of being able to print its results, as more errors arose in printing and proof-reading than in the original calculations

THE DIFFERENCE ENGINE

Charles Babbage and John Herschel were asked by the Royal Astronomical Society to produce new astronomical tables. It was this that caused Babbage to design his calculating machine.

He wanted to mechanize the calculation of a formula such as $x^2 + x + 41$, for different values of x — this was his illustrative example. The core of his idea can be seen in the following table. In the second column are the values of this expression for $x = 0, 1, 2, \ldots, 7$, in the third column are the differences between successive terms of the second (the *first differences*) and in the fourth column are the differences between successive terms of the third column (the *second differences*); here, the second differences are all the same.

x	$x^2 + x + 41$	*first differences*	*second differences*
0	41		
		2	
1	43		2
		4	
2	47		2
		6	
3	53		2
		8	
4	61		2
		10	
5	71		2
		12	
6	83		2
		14	
7	97		

Note that we can reconstruct the values of the function in a steplike fashion from the shaded region containing the first term (41), the initial first difference (2) and the constant second differences (2).

This technique can be applied to any polynomial function, because continuing to take differences eventually yields constant values. Also many functions of interest which are not polynomials (like *sin, cos* and *log*) can be approximated by polynomials.

Charles Babbage

The construction of the difference engine ran into engineering, financial and political difficulties, and construction ended in 1833.

THE ANALYTICAL ENGINE

Babbage wondered whether his difference engine could be made to act upon the results of its own calculations, or as he put it:

The engine eating its own tail.

With this in mind, he designed a new engine, basing its control system on the punched cards used by Jacquard for his automatic loom.

The design for his analytical engine allowed for inputting numbers and holding them in a *store.* The instructions for the operations to be performed on the numbers would be input separately. These operations would be performed in a part of the computer, called the *mill*, and the results would be returned to the store and printed, or used as input for a further calculation, depending on the control instructions. Importantly, the operations to be performed could be made to depend on the result of an earlier calculation.

Ada, Countess of Lovelace, was encouraged in her interest in mathematics by Mary Somerville and Augustus De Morgan.

In her writings on the analytical engine, she described what it could do and how it could be instructed, and gave what is considered to be the first computer program. As she wrote:

Ada, Countess of Lovelace

> *The distinctive characteristic of the Analytical Engine ... is the introduction into it of the principle which Jacquard devised for regulating, by means of punched cards, the most complicated patterns in the fabrication of brocaded stuffs. It is in this that the distinction between the two engines lies. Nothing of the sort exists in the Difference Engine. We may say most aptly that the Analytical Engine weaves algebraical patterns just as the Jacquard loom weaves flowers and leaves.*

Although the analytical engine was never built, modern scholarship is of the view that if it had been constructed, it would have worked as Babbage intended. The name ADA is now given to a programming language developed for the United States Department of Defense.

HAMILTON

William Rowan Hamilton (1805–1865) was a child prodigy who discovered an error in Laplace's treatise on celestial mechanics while still a teenager. He did fundamental theoretical work in mechanics and geometrical optics, using the calculus of variations and building on the principle of least action. He demystified complex numbers and revolutionized algebra with his discovery of quaternions, a non-commutative algebraic system.

Born in Dublin in 1805, Hamilton exhibited a remarkable calculating ability from an early age, while also learning several languages. He began reading Laplace's *Mécanique Céleste* in 1822 and took first place in the entrance examinations for Trinity College the following year, quickly scaling the academic ladder, while becoming professor of astronomy and Astronomer Royal of Ireland in 1827 at the age of 22, just before he graduated.

GEOMETRICAL OPTICS

One of Hamilton's earliest major successes was in the area of geometrical optics, where his theoretical investigations predicted the phenomenon of conical refraction of light in a crystal. This prediction was verified shortly afterwards in 1832 by his Trinity College colleague Humphrey Lloyd, professor of natural philosophy. It caused a sensation, being one of those infrequent occasions when a theoretical investigation predicted previously unknown physical behaviour.

Sir William Rowan Hamilton

In the following diagram are two versions of conical refraction. In both cases the crystal causes a beam of light to be refracted into a hollow cone. The prediction and its verification added further support to the growing acceptance of a wave theory (as opposed to a particle theory) of light.

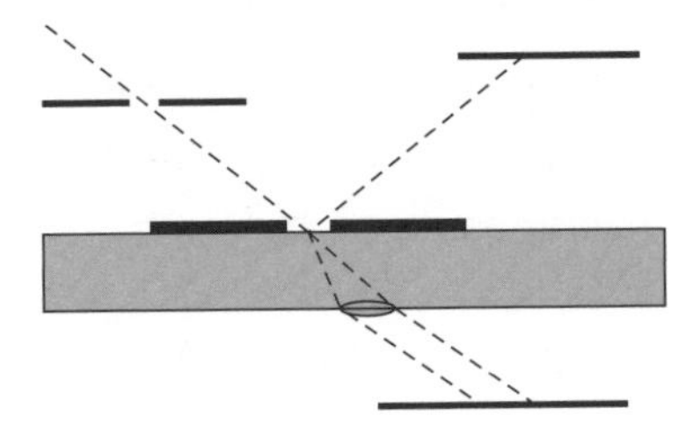

(a) internal conical refraction

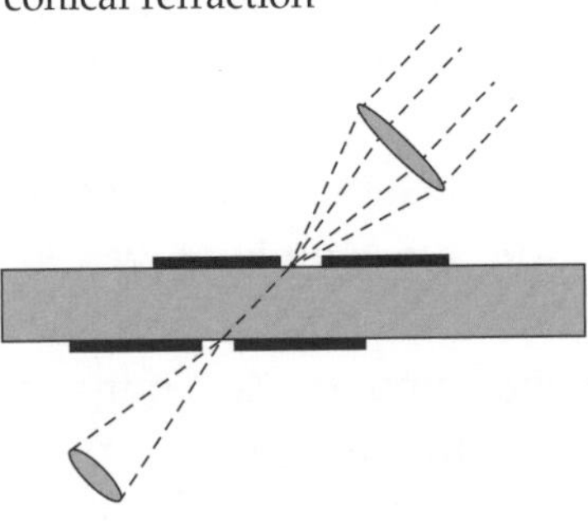

(b) external conical refraction

Hamilton's work was very theoretical and general and his formulation of mechanics, based on the principle of least action, was the one approach to classical mechanics that carried over to quantum mechanics. The *Hamiltonian* is the total energy of the system and is used in both classical and quantum mechanics to discuss the evolution of a system over time.

COMPLEX NUMBERS

For many centuries complex numbers were regarded with suspicion. Euler, who worked with them a great deal, said:

> *Of such numbers we may truly assert that they are neither nothing, nor greater than nothing, nor less than nothing, which necessarily constitutes them imaginary or impossible.*

Even in the early 19th century there was still a great deal of unhappiness about complex numbers, and about so-called 'imaginary' numbers that don't seem to exist. For example, Augustus De Morgan, Professor of Mathematics at University College, London, declared that:

> *We have shown the symbol $\sqrt{-1}$ to be void of meaning, or rather self-contradictory and absurd.*

It was Hamilton who finally provided an explanation of complex numbers that was generally accepted. Recalling the idea of representing each complex number $x + iy$ as a point (x, y) in the plane, he diffused much of the suspicion by proposing that the complex number $a + bi$ should be defined as a pair (a, b) of real numbers. We combine such pairs (a, b) and (c, d) by using the following rules:

Addition:

$$(a, b) + (c, d) = (a + c, b + d);$$

this corresponds to the equation

$$(a + bi) + (c + di) = (a + c) + (b + d)i$$

Multiplication:

$$(a, b) \times (c, d) = (ac - bd, ad + bc);$$

this corresponds to the equation

$$(a + bi) \times (c + di) = (ac - bd) + (ad + bc)i.$$

The pair $(a, 0)$ then corresponds to the real number a, the pair $(0, 1)$ corresponds to the imaginary number i, and we have the equation

$$(0, 1) \times (0, 1) = (-1, 0),$$

which corresponds to the equation $i \times i = -1$.

THE ALGEBRA OF QUATERNIONS

Hamilton then tried to extend his ideas by generalizing the complex numbers to triples in three dimensions. He struggled with this problem for over a decade until he came up with his *quaternions*, each of which consisted of *four* numbers (a, b, c, d), corresponding to an expression of the form

$$a + bi + cj + dk,$$

where $i^2 = j^2 = k^2 = ijk = -1$. However the rule for multiplication is not commutative – the order in which two quaternions are multiplied can make a difference to the answer. In particular,

$$ij = -ji, \quad jk = -kj \quad \text{and} \quad ik = -ki.$$

From these one can deduce that

$$ij = k, \quad jk = i \quad \text{and} \quad ki = j.$$

Hamilton had the idea for his quaternions in 1856 while walking along the Royal Canal in Dublin with his wife. As he recalled:

> *And here there dawned on me the notion that we must admit, in some sense, a fourth dimension of space for the purpose of calculating with triples, ... An electric circuit seemed to close, and a spark flashed forth...*

and he scratched the formulas on Brougham Bridge, now commemorated by a plaque.

The plaque on Brougham Bridge

BOOLE

Although George Boole (1815–1864) contributed to probability, methods of finite differences and differential equations, his greatest achievement was to create an algebra of logic, now called *Boolean algebra.* Not only was this of importance to the development of algebra in the middle of the 19th century, but it remains of current use – for example, in the logical design of digital computing circuits.

George Boole (right) was born in Lincoln, England. A self-taught mathematician, he studied works of Lagrange and Laplace while supporting his family by working as a schoolteacher. He subsequently obtained a position as the first professor of mathematics at the newly established Queen's College Cork in Ireland. Although he worked on differential equations, an overlap of interest with Hamilton, there was little communication between the two men. Indeed, although based in Ireland, Boole had more contact with English mathematicians than with his Irish contemporaries.

BOOLEAN ALGEBRA

In 1854 Boole published his masterpiece, *An Investigation of the Laws of Thought, on Which are Founded the Mathematical Theories of Logic and Probability.* He began:

> *The design of the following treatise is to investigate the fundamental laws of those operations of the mind by which reasoning is performed.*

In his book Boole created a language of symbols and the laws that they satisfy, with a letter representing a set or class of objects.

One of his examples is:

> *If x = class of 'men' and y = class of 'good things', then xy is the class of things belonging to both x and to y (the class of 'good men').*

This particular law of multiplication satisfies the commutative property: $xy = yx$ for all x and y.

For addition, he defined $x + y$ to be the class of all things that belong to either x or y; in his example, $x + y$ is the class of things that are either 'men' or 'good'. He used 0 to represent the class with no members (the empty class) and 1 to represent the universal class, the 'universe of discourse'.

Some of his Boolean algebra laws are:

- $0x = 0$, since the class of things belonging to both 0 and x is the empty class
- $1x = x$, since the class of things belonging to both the universal set and x is x
- $xx = x$, since the class of all things belonging to both x and x is x
- $x + x = x$, since the class of things belonging to either x or x is x

DIFFERENTIAL EQUATIONS

Boole was an important contributor to the theory of differential equations. To illustrate his approach, we consider the differential equation

$$\frac{d^2y}{dx^2} + 2\frac{dy}{dx} - 3y = 0.$$

If D denotes differentiation and D^2 denotes differentiating twice, then we can convert this differential equation into an algebraic one:

$$D^2y + 2Dy - 3y = 0 \text{ or } (D^2 + 2D - 3)y = 0.$$

Factorizing this expression gives

$$(D - 1)(D + 3)y = 0.$$

We now solve $(D - 1)y = 0$ and $(D + 3)y = 0$ and combine the solutions appropriately to solve the original differential equation. The answer is

$$y = Ae^x + Be^{-3x},$$

where A and B are arbitrary constants.

- $(x + y)z = xz + yz$, for all x, y and z
- $x + yz = (x + y)(x + z)$, for all x, y and z.

The ideas in Boolean algebra also arise in probability theory, because the latter may deal with the probability of an outcome that arises from a combination of other possibilities.

John Venn

JOHN VENN

A helpful way of displaying such relationships between classes is to make use of *Venn diagrams,* introduced in 1881 by the Cambridge mathematician John Venn.

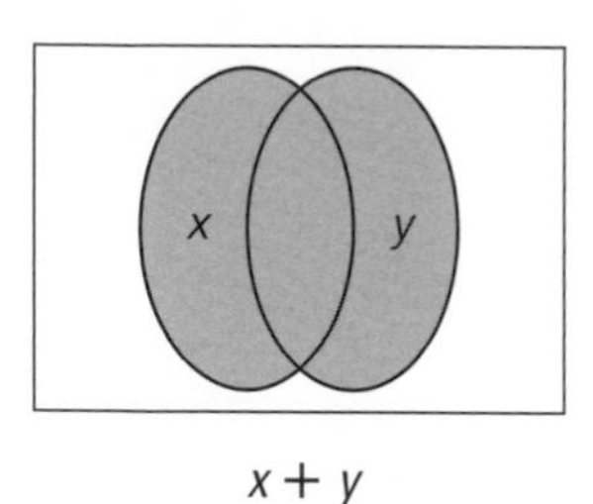

$x + y$

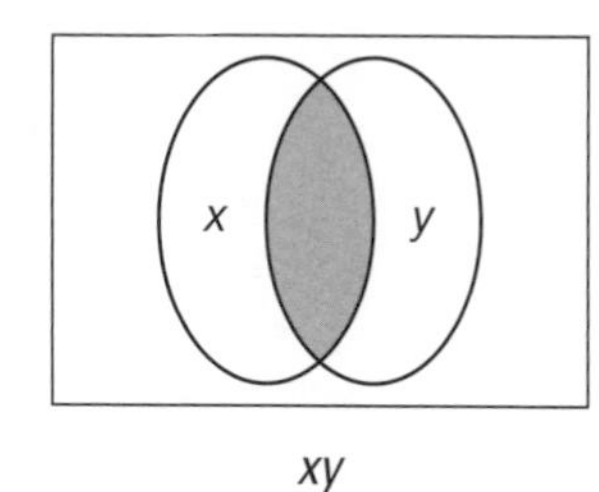

xy

The shaded areas below give the results of adding and multiplying two classes.

We can also use such diagrams to illustrate relationships between classes — for example, the shaded area below represents both

$$x + (yz)$$

and

$$(x + y)(x + z).$$

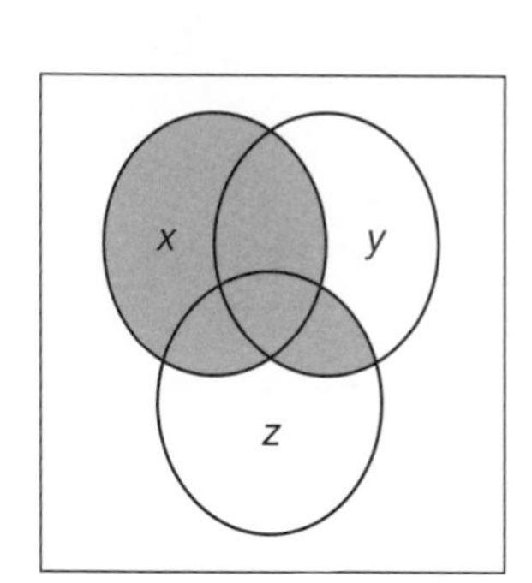

GREEN AND STOKES

George Green (1793–1841) was a pioneering mathematical physicist whose work received little public recognition during his lifetime. He coined the term 'potential', applied it to electricity and magnetism, and is remembered for *Green's theorem* and *Green's function.* George Gabriel Stokes (1819–1903) made major contributions in the areas of hydrodynamics, elasticity, gravity, light, sound, heat, meteorology, solar physics and chemistry. His name is remembered in mathematical physics for *Stokes' theorem* and the *Navier–Stokes equations.*

Green's mill in Sneinton, near Nottingham

Green was born in Nottingham, England and left school at an early age. By trade a miller, he taught himself mathematics by reading works available in his local library, and in 1828 published his most important work, *An Essay on the Application of Mathematical Analysis to the Theories of Electricity and Magnetism.* At age 40 he went to Caius College, Cambridge, to study mathematics, and graduated in 1837.

George Gabriel Stokes

Stokes was born in County Sligo, Ireland, and attended schools in Dublin and Bristol before entering Pembroke College, Cambridge. He graduated as senior wrangler (gaining top marks in the final examinations) in 1841, and became Cambridge's Lucasian Professor of Mathematics in 1849. During his long tenure of this position (for over fifty years until his death), he restored its reputation to the high point it had when occupied by Isaac Newton.

GREEN'S *ESSAY*

Green's stated objective in his *Essay* was to

> *submit to Mathematical Analysis the phenomena of the equilibrium of the Electric and Magnetic Fluids, and to lay down some general principles equally applicable to perfect and imperfect conductors.*

He started by introducing the *potential function,* which he used in his investigation of electricity and magnetism. It is obtained by adding together, for each mass, 'the quantity of electricity that it contains' divided by its distance from a given point.

Green's *Essay* was published by subscription, with about fifty subscribers. It had limited impact until William Thomson obtained a copy in 1845 and republished it with an introduction. Realizing the *Essay*'s importance, Thomson praised its author:

> *His researches ... suggest to the mathematician the simplest and most powerful methods of dealing with problems which, if attacked by the mere force of the old analysis, must have remained forever unsolved.*

8. If X, Y, Z be functions of the rectangular co-ordinates x, y, z, dS an element of any limited surface, l, m, n the cosines of the inclinations of the normal at dS to the axes, ds an element of the bounding line, shew that

$$\iint\left\{l\left(\frac{dZ}{dy}-\frac{dY}{dz}\right)+m\left(\frac{dX}{dz}-\frac{dZ}{dx}\right)+n\left(\frac{dY}{dx}-\frac{dX}{dy}\right)\right\}dS$$
$$=\int\left(X\frac{dx}{ds}+Y\frac{dy}{ds}+Z\frac{dz}{ds}\right)ds,$$

the differential coefficients of X, Y, Z being partial, and the single integral being taken all round the perimeter of the surface.

The first appearance in print of Stokes' theorem, as an examination question in 1854!

GREEN'S AND STOKES' THEOREM

The *Fundamental theorem of the calculus*, representing differentiation and integration as inverse processes, can be written:

$$\int_a^b df = f(b) - f(a).$$

It may be thought of as relating the behaviour of one function inside the interval from a to b to the values of another one at the endpoints (or boundary) a and b. The theorems carrying the names of Green and Stokes (although they were not the first to discover them) generalize this idea to two and three dimensions.

In particular, Green's theorem (below) relates the values of one function inside the area D to that of another one on its boundary C.

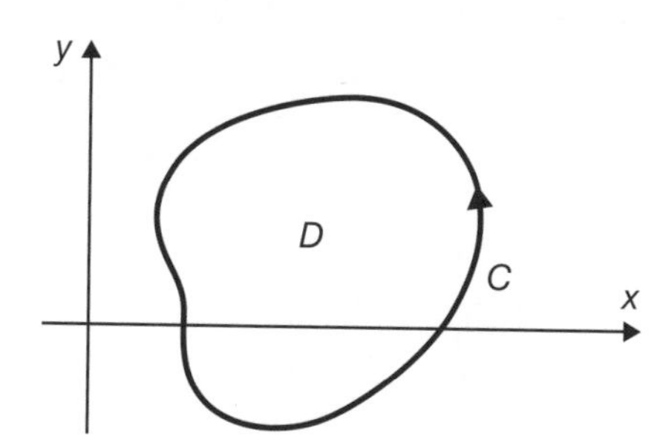

STOKES AND HYDRODYNAMICS

In later years, Stokes explained why he took up the study of hydrodynamics:

> *I thought I would try my hand at original research; and, following a suggestion made to me by Mr. Hopkins* [a famous Cambridge tutor] *while reading for my degree, I took up the subject of Hydrodynamics, then at rather a low ebb in the general reading of the place, notwithstanding that George Green, who had done admirable work in this and other departments, was resident in the University till he died.*

At the 1846 meeting of the British Association for the Advancement of Science, Stokes reported on hydrodynamics. His perceptive survey enhanced his reputation and showed his familiarity with the work of Lagrange, Laplace, Fourier, Poisson and Cauchy, as well as that of Green. As he said in his report:

> *The fundamental hypothesis on which the science of hydrostatics is based may be considered to be, that the mutual action of two adjacent portions of a fluid at rest is normal to the surface which separates them ... and thus the hypothesis above-mentioned may be considered as the fundamental hypothesis of the ordinary theory of hydrodynamics, as well as hydrostatics.*

After 1850 his output of academic publications slowed. This was due partly to his role in academic administration — for example, he became Physical Sciences Secretary of the Royal Society — but also to the time and effort that he spent in corresponding with colleagues, encouraging them and commenting on their work, and communicating their results.

THOMSON AND TAIT

Peter Guthrie Tait

William Thomson (1824–1907), later known as Lord Kelvin, was a dominant figure in Victorian science with contributions to mathematics, physics and engineering, particularly in the areas of electricity and magnetism. A leading figure in the creation of thermodynamics, the area of physics concerned with heat and energy, he was instrumental in the laying of the first transatlantic telegraph cable. Peter Guthrie Tait (1831–1901) carried out research in a wide range of topics: quaternions, knot theory, atmospheric and meteorological phenomena, thermodynamics, aerodynamics and kinetic theory, as well as the colouring of maps. In the 1860s, Thomson and Tait collaborated on a highly influential book on natural philosophy.

Thomson was born in Belfast, Ireland. He was educated at Glasgow and Cambridge Universities, and was appointed Professor of Natural Philosophy at Glasgow University at the age of 22. He remained at Glasgow until his death and was buried in Westminster Abbey alongside Isaac Newton.

Tait was born in Dalkeith, Scotland, and was educated at Edinburgh and Cambridge Universities. He became Professor of Mathematics at Queen's College, Belfast, in 1854 and took up the Chair of Natural Philosophy at Edinburgh in 1860, which he then held for more than 40 years. He collaborated with Thomson, and also with Maxwell and Hamilton.

TREATISE ON NATURAL PHILOSOPHY

Probably the most influential legacy of the collaboration between Thomson and Tait was the production in 1867 of *The Treatise on Natural Philosophy.* They started to work on it in 1861, shortly after they met, and it proved to be highly influential in identifying and placing the conservation of energy at the heart of its approach, although in scope it fell far short of its authors' original intentions.

The two men had very different natures. Thomson frequently went travelling, while Tait did not leave Scotland after 1875. Tait could also be argumentative, with bitter disputes (for example) with Heaviside and Gibbs over the relative merits of vectors and quaternions — but it was he who drove the collaboration towards publication, berating, cajoling and coaxing Thomson to keep to deadlines. Tait's frustration is illustrated in a letter he wrote to Thomson in June 1864, about halfway through the collaboration:

> *I am getting quite sick of the great Book ... if you send only scraps and these at rare intervals, what can I do? You have not given*

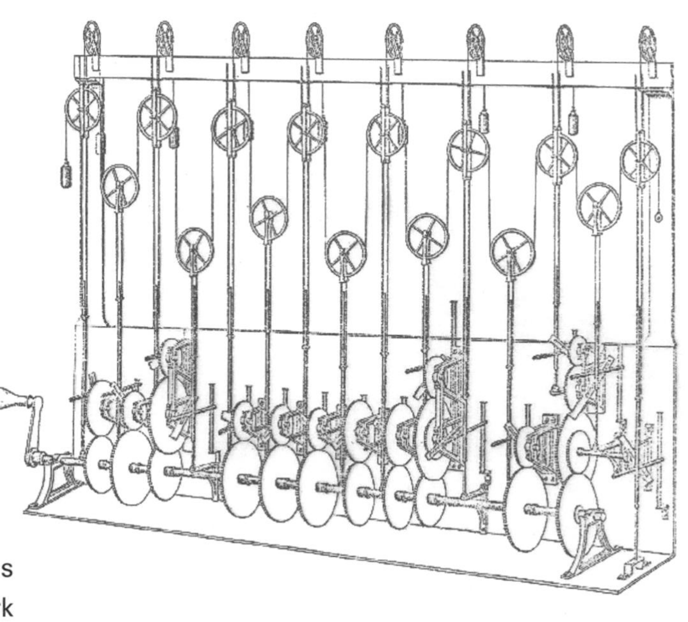
A tide-predicting machine

me even a hint as to what you want done in our present chapter about statics of liquids and gases!

The treatise was universally abbreviated to T & T′, and Thomson and Tait used this in their extensive correspondence. As one of its formulas was $dp/dt = jcm$, their close friend James Clerk Maxwell became known as *dp*/*dt*! Maxwell and Tait had been friends since they first went to school at Edinburgh Academy at the age of 10, and on from there to Edinburgh University and Cambridge.

Thomson and Tait had three main purposes for the project that resulted in T&T′:

- to provide appropriate affordable textbooks to back up their lectures
- to nurture physical intuition and mitigate against relying on mathematical manipulation
- to base their natural philosophy on the conservation of energy and extremum principles, replacing 'Newton's Principia of force with a new Principia of energy and extrema'.

Their treatise was well received. Maxwell's view of their achievement was:

The credit of breaking up the monopoly of the great masters of the spell, and making all their charms familiar to our ears as household words, belongs in great measure to Thomson and Tait. The two northern wizards were the first who, without compunction or dread, uttered in their mother tongue the true and proper names of those dynamical concepts which the magicians of old were wont to invoke only by the aid of muttered symbols and inarticulate equations. And now the feeblest among us can repeat the words of power and take part in dynamical discussions which but a few years ago we should have left for our betters.

Lord and Lady Kelvin at the Coronation of King Edward VII

CONTINUOUS CALCULATING MACHINES

The second edition of T & T′ features a marvellous discussion on continuous calculating machines, bringing together previous work by Thomson and his brother James. There are machines for solving simultaneous equations, integrating the product of two given functions, and finding the solutions of linear second-order differential equations with variable coefficients. A machine for predicting tides computed the depth of water over a period of years, for any port for which the 'tidal constituents have been found from harmonic analysis of tide-gauge observations' — that is, from the coefficients of the Fourier series representing the rise and fall of the tide.

MAXWELL

James Clerk Maxwell (1831–1879) is widely considered to be one of the most important mathematical physicists of all time, after only Newton and Einstein. Foremost among his contributions to science was the formulation of the theory of electromagnetism, with light, electricity and magnetism all shown to be manifestations of the electromagnetic field. He also made major contributions to the theory of colour vision and optics, the kinetic theory of gases and thermodynamics, and understanding the dynamics and stability of Saturn's rings.

Maxwell was born in Edinburgh, Scotland, entering the University there when he was 16. In 1850 he moved to Cambridge University, and six years later returned to Scotland to the Chair of

James Clerk Maxwell

Michael Faraday delivering a Christmas Lecture at the Royal Institution in 1856

Natural Philosophy at Marischal College, Aberdeen, being made redundant when the College was amalgamated into Aberdeen University.

He was subsequently professor at King's College, London, and first Cavendish Professor of Physics at the Cavendish Laboratory in Cambridge where he designed the laboratory and contributed to the purchase of its equipment; he also edited and commented on Cavendish's papers on electricity.

ELECTROMAGNETISM

The English physicist Michael Faraday and the Danish physicist Hans Christian Oersted made fundamental discoveries in electricity and magnetism. Among these were:

- the conversion of electrical energy into mechanical energy, since electrical currents create magnetic fields
- the conversion of mechanical energy into electrical energy, since a moving magnet induces an electric current in a wire.

These observations were crucial to Maxwell's work, a point that Maxwell repeatedly stressed. His main achievement was to formulate

Maxwell's equations

$$\text{curl } \mathbf{H} = j + \partial\mathbf{D}/\partial t$$
$$\text{div } \mathbf{B} = 0$$
$$\text{curl } \mathbf{E} = -\partial\mathbf{B}/\partial t$$
$$\text{div } \mathbf{D} = \rho$$

mathematically Faraday's work on electric and magnetic lines of force. In a few relatively simple equations, Maxwell captured the behaviour of electric and magnetic fields and their interaction. His calculations showed that the speed of propagation of an electromagnetic field is approximately that of light, and he wrote:

> *We can scarcely avoid the conclusion that light consists in the transverse undulations of the same medium which is the cause of electric and magnetic phenomena.*

His *Treatise on Electricity and Magnetism* was published in 1873. The influence of his work was profound. Einstein enthused:

> *Since Maxwell's time, physical reality has been thought of as represented by continuous fields, and not capable of any mechanical interpretation. This change in the conception of reality is the most profound and the most fruitful that physics has experienced since the time of Newton*

while the distinguished physicist Richard Feynman predicted:

> *From a long view of the history of mankind — seen from, say, ten thousand years from now — there can be little doubt that the most significant event of the 19th century will be judged as Maxwell's discovery of the laws of electrodynamics.*

Tait and Thomson were friends and correspondents of Maxwell, and Maxwell thanked them in the introduction to his *Treatise.*

MAXWELL'S DEMON

In December 1867 Maxwell wrote to Tait outlining a thought experiment connected with the second law of thermodynamics. He imagined a container, filled with a gas, with a partition dividing it in two. The molecules of the gas are moving with different speeds. There is a small hole in the partition which a 'being' can open or close, in order to allow swifter molecules to pass only from the left side to the right. With no expenditure of work, the being increases the temperature in the right side and lowers the temperature in the left side, contradicting the second law of thermodynamics. William Thompson (Lord Kelvin) was the first to use the word 'demon' for Maxwell's concept.

A LECTURE ON THOMSON'S GALVANOMETER

Delivered to a single pupil in an alcove with drawn curtains

The lamp-light falls on blackened walls,
And streams through narrow perforations;
The long beam trails o'er pasteboard scales,
With slow-decaying oscillations.
Flow, current! flow! set the quick light-spot flying!
Flow, current! answer, light-spot! flashing, quivering, dying.

O look! how queer! how thin and clear,
And thinner, clearer, sharper growing,
This gliding fire, with central wire
The fine degrees distinctly showing.
Swing, magnet! swing! advancing and receding:
Swing, magnet! answer, dearest, what's your final reading?

O love! you fail to read the scale
Correct to tenths of a division;
To mirror heaven those eyes were given,
And not for methods of precision.
Break, contact! break! set the free light-spot flying!
Break, contact! rest thee, magnet! swinging, creeping, dying.

$\frac{dp}{dt}$

Maxwell's sense of fun is shown in this poem to Thomson's galvanometer (an instrument for measuring current)

KIRKMAN

It is unfortunate when mathematical concepts become credited to the wrong person. This happened twice to the Revd. Thomas Penyngton Kirkman (1806–1895), a country parson in the English county of Lancashire, who made significant contributions to the study of triple systems, polyhedra, groups and knots.

Thomas Kirkman was born in Bolton, in Lancashire. After working for a few years in his father's office, he went to Trinity College, Dublin, where he studied mathematics and other subjects. After graduating, he entered the church, eventually becoming Rector of the Lancashire parish of Croft-with-Southworth, a position that he then held for fifty-two years.

His parochial duties consumed little of his energies, and he spent his spare time enjoying family life with his wife and seven children and writing papers on a wide range of pure mathematical topics. Living so far from London and mixing rarely with other mathematicians, he tended to make up his own terminology (much of it incomprehensible) and found it difficult to have his researches fully accepted.

TRIPLE SYSTEMS

In the *Lady's and Gentleman's Diary* of 1846, the editor asked when it is possible to arrange the numbers from 1 to n in triples so that any two numbers appear together in exactly one triple; for example, when $n = 7$ we can arrange the triples vertically as follows,

$$\begin{array}{ccccccc} 1 & 2 & 3 & 4 & 5 & 6 & 7 \\ 2 & 3 & 4 & 5 & 6 & 7 & 1 \\ 4 & 5 & 6 & 7 & 1 & 2 & 3 \end{array}$$

and when $n = 9$ we have the arrangement

$$\begin{array}{cccccccccccc} 1 & 1 & 1 & 1 & 2 & 2 & 2 & 3 & 3 & 3 & 4 & 7 \\ 2 & 4 & 5 & 6 & 4 & 5 & 6 & 4 & 5 & 6 & 5 & 8 \\ 3 & 7 & 9 & 8 & 9 & 8 & 7 & 8 & 7 & 9 & 6 & 9 \end{array}$$

Simple counting arguments show that such arrangements can be possible only when n has the form $6k + 1$ or $6k + 3$, for some integer k — that is, n is one of the numbers

7, 9, 13, 15, 19, 21, 25, 27, … .

Kirkman became interested in such systems, and showed how to construct them for all such values of n.

He further investigated whether it is possible to arrange the triples into blocks, each containing all the numbers from 1 to n. For example, when $n = 9$ one can arrange the triples as follows:

$$\begin{array}{ccc|ccc|ccc|ccc} 1 & 4 & 7 & 1 & 2 & 3 & 1 & 2 & 3 & 1 & 2 & 3 \\ 2 & 5 & 8 & 4 & 5 & 6 & 6 & 4 & 5 & 5 & 6 & 4 \\ 3 & 6 & 9 & 7 & 8 & 9 & 8 & 9 & 7 & 9 & 7 & 8 \end{array}$$

In the *Lady's and Gentleman's Diary* of 1850, he asked for a similar arrangement of triples when $n = 15$:

Fifteen young ladies in a school walk out three abreast for seven days in succession: it is required to arrange them daily, so that no two shall walk twice abreast.

This puzzle became known as *Kirkman's schoolgirls problem,* and his solution to it appeared in the 1851 *Diary.*

Two years later, the celebrated Swiss geometer Jakob Steiner wrote a short note in a German journal asking when

The Revd. Thomas Penyngton Kirkman

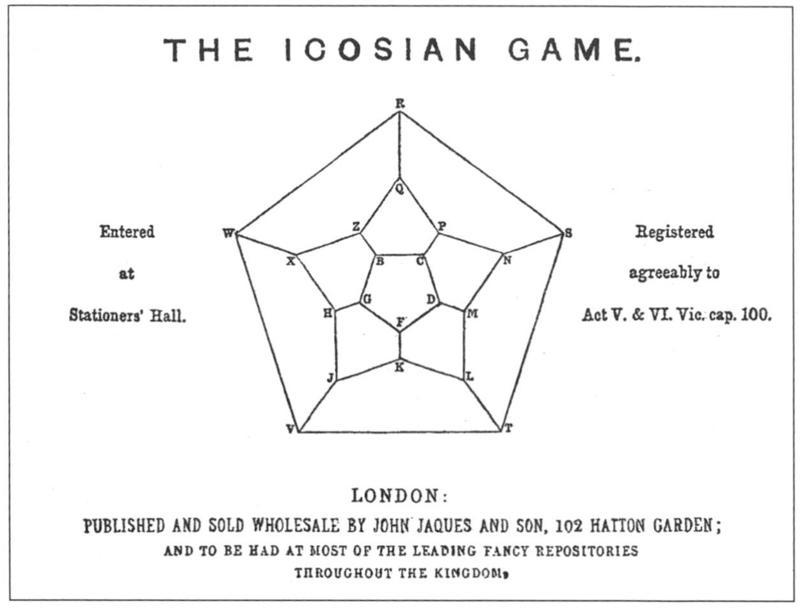

such systems can be constructed, completely unaware that Kirkman had published a solution to this problem six years earlier. Because of Steiner's prominence, such systems are now called *Steiner triple systems.*

POLYHEDRA

Another area in which Kirkman did not receive the credit due to him was in the study of polyhedra (or 'polyedra' as he called them). In 1856, arising from his work on algebraic systems, William Rowan Hamilton had been led to investigate cyclic routes on a dodecahedron. Fascinated by these, he marketed a puzzle, called the *Icosian game,* in which he labelled the vertices with the twenty consonants of the alphabet, and challenged his readers to find 'A Voyage round the World', visiting each city from B (Brussels) to Z (Zanzibar) exactly once before returning to the beginning (see above).

A year earlier than Hamilton, Kirkman had written extensively about cyclic routes on polyhedra in general (and not just dodecahedra). These cyclic routes are now known as *Hamiltonian cycles,* although the priority clearly belongs to Kirkman.

GROUPS

A *group* is an algebraic object that consists of a set of elements and a way of combining them in pairs so as to satisfy certain specified rules.

Examples of groups include:

- adding integers to give other integers
- multiplying positive numbers to give other positive numbers
- combining the symmetries of a cube
- combining all the permutations of a given collection of objects.

Group theory arose originally from permuting the solutions of equations, and Lagrange and Cauchy had been early workers in the area, but it was the work of Galois that really initiated the subject. Later writers on group theory included Cayley and Kirkman.

In 1857 the French Academy of Sciences proposed a prize competition on an investigation into groups. The three entrants included Kirkman but, to his disgust, the prize was not awarded.

KNOTS

Towards the end of his life, Kirkman became interested in some researches of Tait into the study of knots. Tait had constructed tables of all the different types of knot with up to seven crossings, and Kirkman collaborated with him to find all the knots with eight, nine and ten crossings. In recent years, knot theory has become a very active area of study.

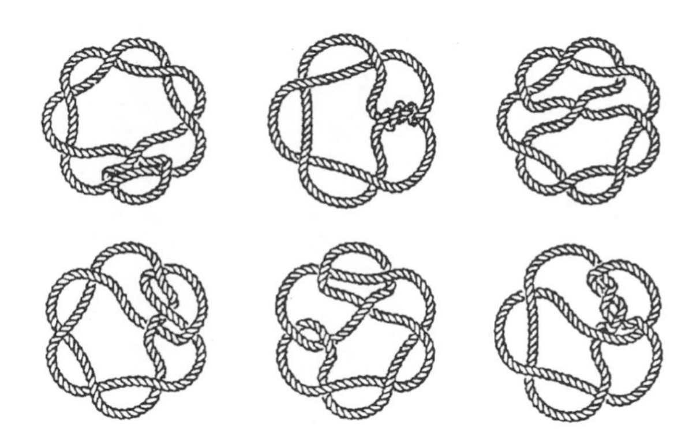

Some knots with eight crossings

CAYLEY AND SYLVESTER

In his book *Men of Mathematics* (1937), E. T. Bell dubbed Arthur Cayley (1821–1895) and James Joseph Sylvester (1814–1897) the 'invariant twins', for their joint contributions to the algebraic theory of invariants. Their lives make an interesting comparison: both ran foul of Cambridge's religious rules, both worked for a while in London, and together they transformed algebra in Britain. Yet temperamentally they were poles apart, Cayley's unruffled and methodical personality often contrasting with that of his more impetuous and disorganized friend.

CAYLEY

From an early age, Arthur Cayley developed a remarkable ability for mathematics. At the age of 14 he enrolled as a day pupil at King's College, London, and progressed from there to Trinity College, Cambridge. There he enjoyed a glittering undergraduate academic career, emerging top of his year and winning the coveted Smith's prize in mathematics.

With such a spectacular start to his career, he was naturally awarded a fellowship at Trinity College, but in those days college fellows were required to train for the priesthood and Cayley had no wish to do so. He left Trinity to go to Gray's Inn, London, and train as a lawyer. Shortly after arriving there he met Sylvester, and thus began their remarkable friendship and mathematical collaboration.

Arthur Cayley

During his seventeen years as a successful barrister in London, Cayley wrote over two hundred mathematical papers, including some of his most important contributions to the subject — in particular, initiating the algebra of matrices and invariant theory (the study of algebraic expressions left unchanged by certain transformations).

In 1863 Cambridge University founded the Sadleirian Chair of Pure Mathematics, with no religious requirements. Cayley was duly appointed and returned to his *alma mater,* where he spent the rest of his life.

One of the most prolific mathematicians of all time, Cayley produced almost one thousand research papers at a remarkable rate in a wide variety of topics – from algebra and geometry to analysis and astronomy.

SYLVESTER

Sylvester also showed early promise in mathematics, attending Augustus De Morgan's mathematics classes at University College, London, at the age of 14. Although not an Orthodox Jew, his faith was important to him, and in his early life he was subjected to insults and prejudice. Although permitted to study at Cambridge, he could not receive his degree until the rules were changed in 1871. Moreover, although placed second in the final examinations, he could not hold a fellowship at either Oxford or Cambridge.

James Joseph Sylvester

TREES AND CHEMISTRY

In addition to their algebraic pursuits, both Cayley and Sylvester became involved with the study of *tree structures.* Resembling a family tree, a *mathematical tree* consists of a number of points, joined up in such a way that no cycles appear.

John

Joe Jean Jane Jill

Jenny Kenny Bill Ben

Family tree

Mathematical tree

Cayley was involved with counting trees, and developed iterative methods for finding the number of different trees with any given number of points — for example, there are just three types of tree with five points.

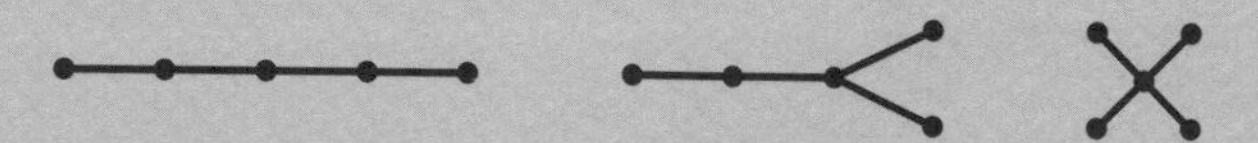

In the meantime, Sylvester had become interested in chemistry and noted the tree structure of certain types of chemical molecule, such as the paraffins (alkanes) C_nH_{2n+2} and the alcohols $C_nH_{2n+1}OH$. Cayley then developed tree-counting methods for enumerating such molecules.

```
    H  H  H              H  H  H
    |  |  |              |  |  |
H—C—C—C—H          H—C—C—C—O—H
    |  |  |              |  |  |
    H  H  H              H  H  H
```

propane (C_3H_8)

propyl alcohol (C_3H_7OH)

But Sylvester desired an academic career that would enable him to carry out his mathematical researches. He was appointed to professorships of natural philosophy in London, and mathematics at the University of Virginia, but neither proved satisfactory, and in the mid-1840s he was back in London with no academic position. He became an actuary at the Equity and Law Life Assurance Society, and while in London met and worked with Arthur Cayley. In 1855 he was appointed professor of mathematics at the Royal Military Academy at Woolwich, where he remained for fifteen years until military regulations forced him to retire at the age of 55.

His days of regular employment seemed to be over and he pursued other interests, such as singing and writing poetry. But in 1876 he was head-hunted as the first mathematics professor at the newly founded Johns Hopkins University in Baltimore, USA. There he spent seven happy and productive years, working on his own research, training others to be professional mathematicians, and building a research school of a type known in Continental Europe but previously unknown in Great Britain or the United States.

In 1883, at the age of 69, Sylvester returned to England to embark on his final career, when he was appointed Savilian Professor of Geometry at Oxford University, a position that he held until failing eyesight forced him to give up the post.

CHEBYSHEV

Pafnuty Chebyshev

Pafnuty Chebyshev (1821–1894) taught at the University of St Petersburg and founded the St Petersburg School of Mathematics. He is remembered mainly for his work on orthogonal functions and probability, and for an important contribution to the proof of the prime number theorem. He also worked on quadratic forms and integrals and studied theoretical mechanics and linkages.

Chebyshev was one of the most eminent Russian mathematicians of the 19th century. Born in Okatovo, Western Russia, he attended Moscow University in 1837. After his studies he moved to St Petersburg as professor of mathematics, and remained there until he retired.

APPROXIMATION

Chebyshev was interested in mechanics and machines and believed in the beneficial effects of the interaction between theory and practice. In 1856 he wrote:

> *The closer mutual approximation of the points of view of theory and practice brings most beneficial results, and it is not exclusively the practical side that gains; under its influence the sciences are developing, in that this approximation delivers new objects of study or new aspects in subjects long familiar.*

His work on the approximation of functions by *Chebyshev polynomials* was partly motivated by his interest in the theory of machines. Using integration, he defined a concept for two polynomials that is analogous to the angle between two intersecting lines. We say that two lines are *orthogonal* if they meet at right angles, and with his more general concept of an angle, all the Chebyshev polynomials can be thought of as orthogonal to each other. He developed this idea in his general theory of orthogonal polynomials.

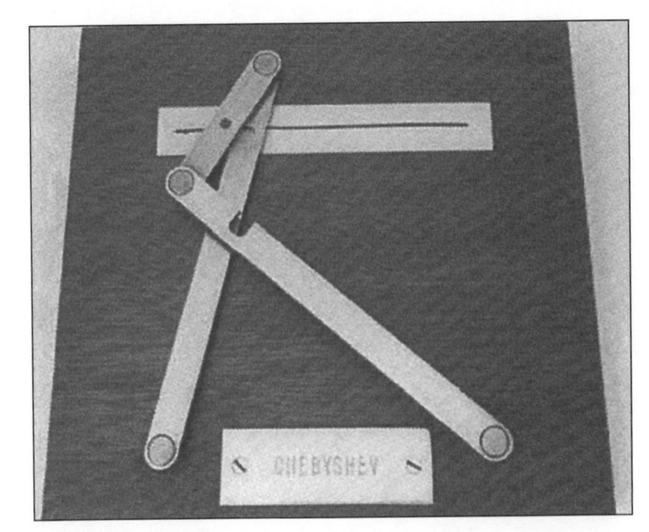

Chebyshev's linkage that draws an approximation to a straight line, as used (for example) for the motion of an engine piston

THE PRIME NUMBER THEOREM

Although the list of prime numbers continues for ever, the primes themselves are irregularly distributed. In the words of the number-theorist Don Zagier:

> ***The prime numbers belong to the most arbitrary objects studied by mathematicians: they grow like weeds, seeming to obey no other law than that of chance, and nobody can predict where the next one will be.***

In particular, there seem to be 'twin primes' (pairs differing by 2, such as 29 and 31, or 107 and 109) as far as we go, although this has never been proved, and yet we can also find arbitrarily long strings of non-primes. However, the primes do thin out fairly regularly, the further up we go. As Don Zagier paradoxically added:

> ***The prime numbers exhibit stunning regularity, there are laws governing their behaviour, and they obey these laws with almost military precision.***

What are these laws? Around 1792, the 15-year-old Gauss constructed extensive tables of prime numbers up to three million, and noticed that the density of primes near each number n is about $1 / \log_e n$; this is equivalent to saying that if $P(n)$ denotes the number of primes up to n (so that $P(10) = 4$, the primes being 2, 3, 5, 7), then $P(n)$ behaves very much like $n / \log_e n$, when n becomes large – more precisely, the ratio of $P(n)$ and $n / \log_e n$ has limit 1 as n tends to infinity; this is the *Prime number theorem.*

But Gauss and his contemporaries were unable to prove this. Around 1851, Chebyshev proved that *if* this ratio tends to a limit, *then* this limit is 1. But the Prime number theorem was not fully proved until 1896 – by the Frenchman Jacques Hadamard and the Belgian Charles-Jean de la Vallée Poussin, independently.

PROBABILITY

Chebyshev is considered the intellectual father of the long and distinguished series of Russian probabilists who contributed to the subsequent development of the subject.

A probability distribution is a curve that describes the variation in a set of results. The probabilities are then given by areas under the curve — for example, in the figure below, the probability of an observation between *a* and *b* is the shaded area between these values. An important example is the *normal distribution,* also called the *Gaussian distribution* or (because of its shape) the *bell curve.*

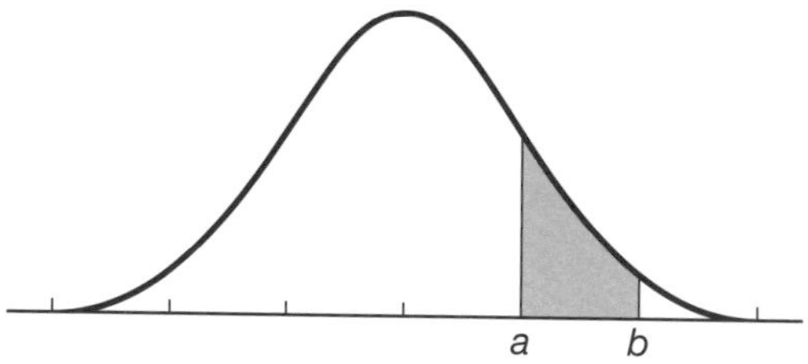

The *average* (or *mean* or *expected*) *value* of a probability distribution can be calculated by integration, essentially by adding together the product of each observation by its probability. *Chebyshev's inequality* places an upper limit on the probability that the observations differ from the average. It is a powerful result and, as the mathematician Andrey Kolmogorov wrote:

> *The principal meaning of Chebyshev's work is that, through it, he always aspired to estimate exactly in the form of inequalities ... the possible deviations from limit regularities. Further, Chebyshev was the first to estimate clearly, and make use of, such notions as 'random quantity' and its 'expectation (mean) value'.*

NIGHTINGALE

Florence Nightingale (1820–1910), the 'lady with the lamp' who saved lives during the Crimean War, was also a fine statistician who collected and analysed mortality data from the Crimea and displayed them on her 'polar diagrams', a forerunner of the pie chart. Her work was strongly influenced by that of the Belgian statistician Adolphe Quetelet.

Florence Nightingale showed an early interest in mathematics — at the age of 9 she was displaying data in tabular form, and by the time she was 20 she was receiving tuition in mathematics, possibly from James Joseph Sylvester.

Nightingale regarded statistics as 'the most important science in the world' and used statistical methods to support her efforts at administrative and social reform. She was the first woman to be elected a Fellow of the Royal Statistical Society and an honorary foreign member of the American Statistical Association.

Florence Nightingale

STATISTICAL INFLUENCE

By 1852 Nightingale had established a reputation as an effective administrator and project manager. Her work on the professionalization of nursing led to her accepting the position of 'Superintendent of the female nursing establishment in the English General Military Hospitals in Turkey' for the British troops fighting in the Crimean war. She arrived in 1854 and was appalled at what she found there. In attempting to change attitudes and practices she made use of pictorial diagrams for statistical information, developing her *polar area graphs.*

The graphs have twelve sectors, one for each month, and reveal changes over the year in the deaths from wounds obtained in battle, from diseases, and from other causes. They showed dramatically the extent of the needless deaths amongst the soldiers during the Crimean war, and were used to persuade medical and other professionals that deaths could be prevented if sanitary and other reforms were made.

On her return to London in 1858, she continued to use statistics to inform and influence public health policy. She urged the collection of the same data, across different hospitals, of:

- the number of patients in hospital
- the type of treatment, broken down by age, sex and disease
- the length of stay in hospital
- the recovery rate of patients.

She argued for the inclusion in the 1861 census of questions on the number of sick people in a household, and on the standard of housing, as she realised the important relationship between health and housing. In another initiative she tried to educate members of the government

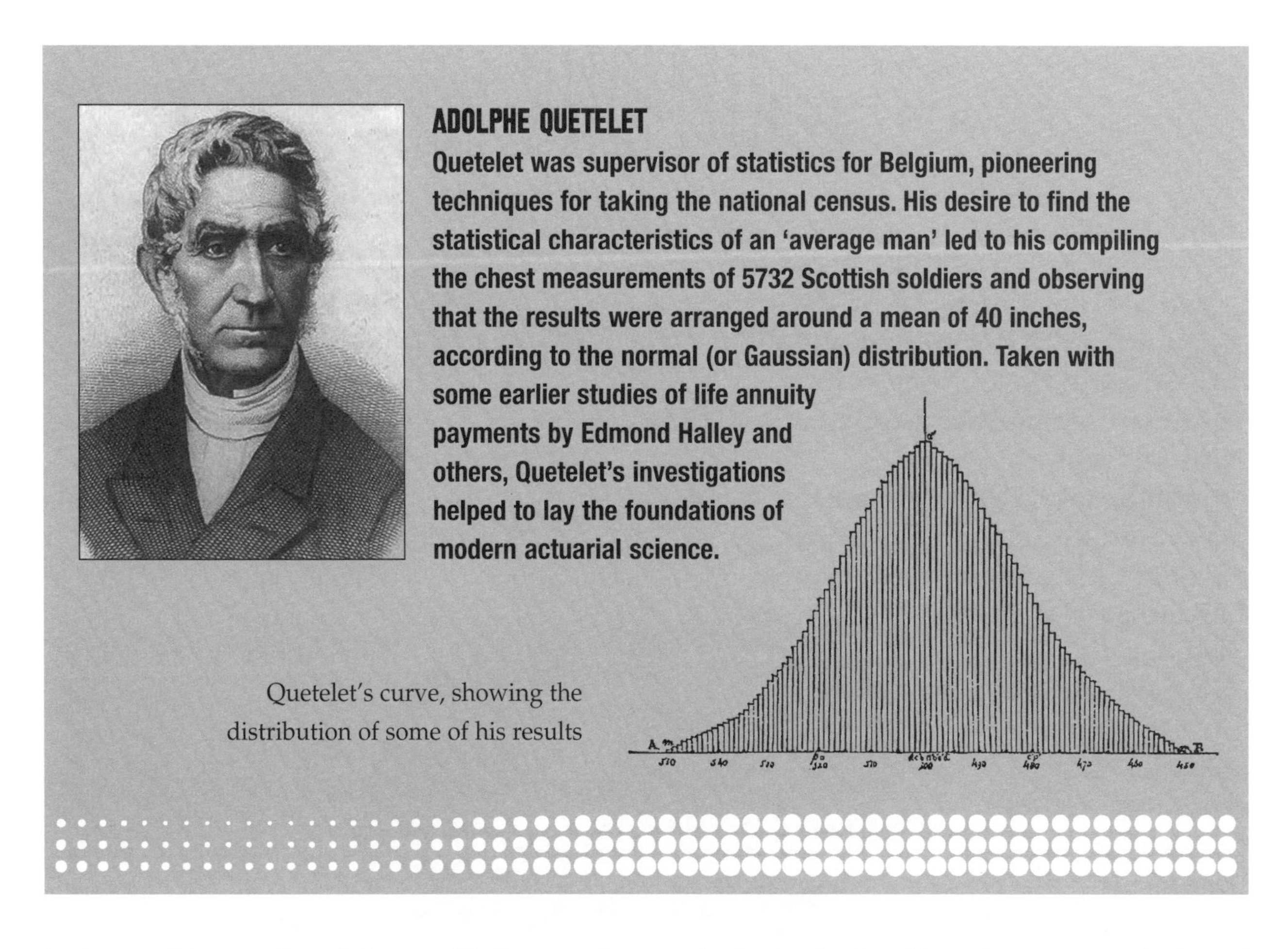

ADOLPHE QUETELET

Quetelet was supervisor of statistics for Belgium, pioneering techniques for taking the national census. His desire to find the statistical characteristics of an 'average man' led to his compiling the chest measurements of 5732 Scottish soldiers and observing that the results were arranged around a mean of 40 inches, according to the normal (or Gaussian) distribution. Taken with some earlier studies of life annuity payments by Edmond Halley and others, Quetelet's investigations helped to lay the foundations of modern actuarial science.

Quetelet's curve, showing the distribution of some of his results

in the usefulness of statistics, and influence the future by establishing the teaching of the subject in the universities.

For Nightingale the collection of data was only the beginning. Her subsequent analysis and interpretation was crucial and led to medical and social improvements and political reform, all with the aim of saving lives.

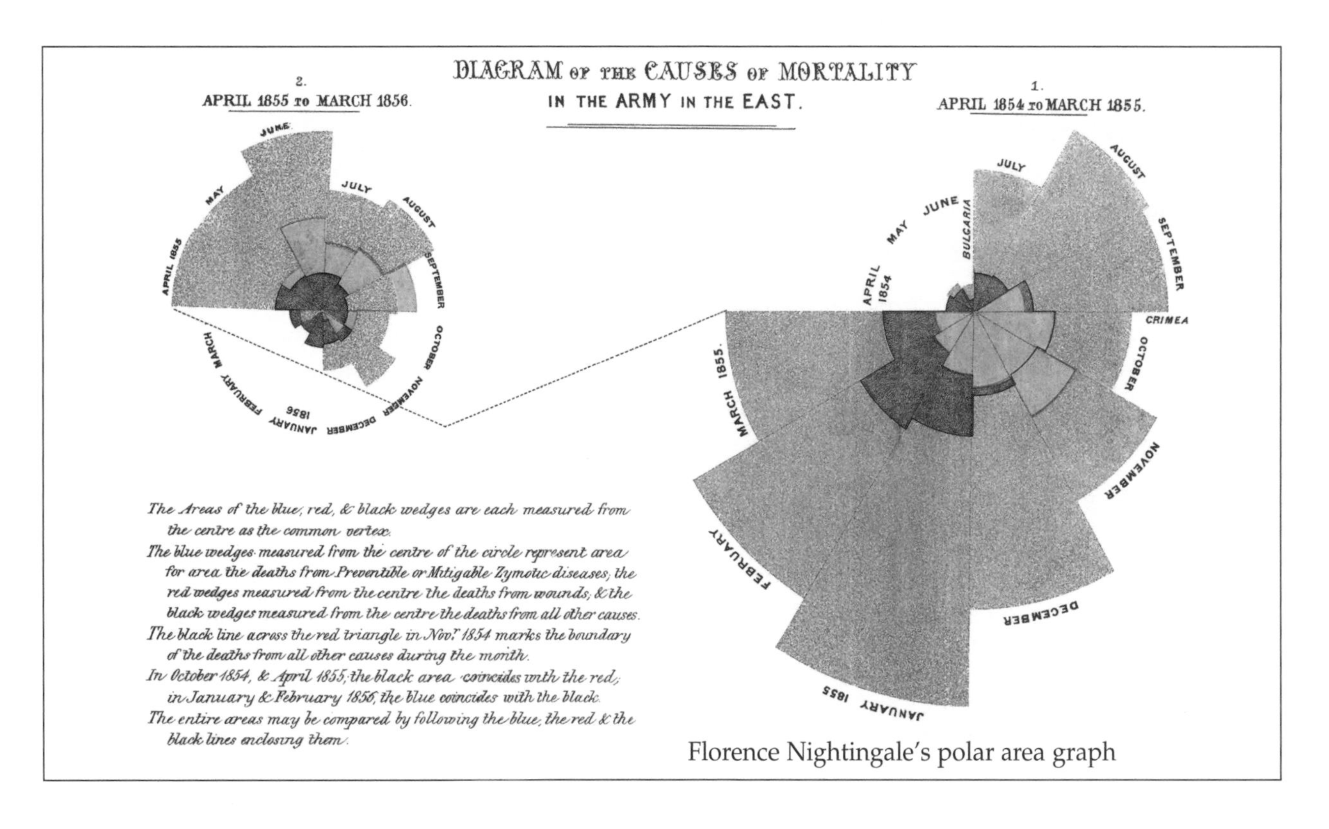

Florence Nightingale's polar area graph

RIEMANN

In many areas the work of Bernhard Riemann (1826–1866) had as much influence as that of any other 19th-century mathematician. With his remarkable combination of geometrical reasoning and physical insight, he developed the general theory of functions of a complex variable, using his 'Riemann surfaces' as a bridge between analysis and geometry, while developing the theory of integration and the convergence of series. In another direction, he obtained a remarkable generalization of the idea of 'geometry', both Euclidean and non-Euclidean; years later, one of his geometries proved to be the natural setting for Einstein's theory of relativity. In number theory he left us with what many consider to be the most important unsolved problem in mathematics.

Riemann was born in Breselenz in Northern Germany and studied at the University of Göttingen, obtaining his doctorate there in 1851. Gauss wrote that Riemann's thesis exhibited:

> *a creative, active, truly mathematical mind, and of a gloriously fertile originality.*

In 1859, Riemann was appointed to Gauss's former position as professor of mathematics at Göttingen, succeeding the algebraist and number-theorist Lejeune Dirichlet. Riemann held this position for seven years, until his early death at the age of 40.

RIEMANN'S GEOMETRY

Riemann's work in complex analysis was the true beginning of topology, the area of geometry concerned with those properties of space that are unchanged by continuous deformation.

Riemann was also interested in the geometry of higher dimensions. Although we cannot visualize dimensions larger than three, we can still investigate them mathematically. Any point in the two-dimensional plane can be represented by two coordinates (a, b), and similarly we can represent any point in three-dimensional space by three coordinates (a, b, c).

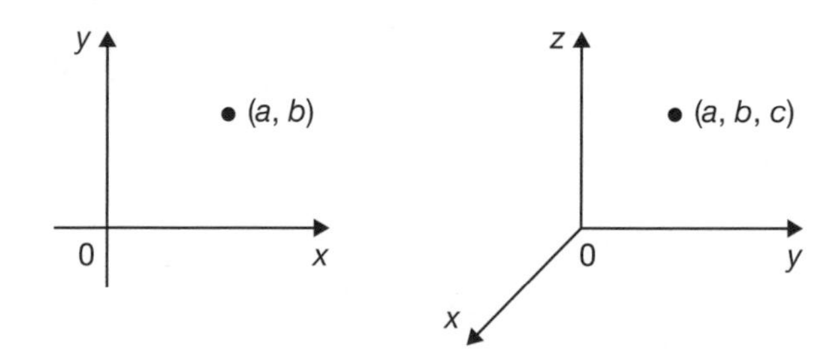

Analogously, we can represent any point in four-dimensional space by four coordinates (a, b, c, d), and similarly for five, six, and higher dimensions. We can then work out lengths and angles in these higher dimensions — just as before, but less easy to visualize.

Riemann also studied the way that surfaces can 'curve' inwards or outwards (like a globe or a cooling tower) and proposed generalized ideas of distance for such surfaces, not only in three dimensions, but also for their higher-dimensional analogues (called *manifolds*). By disregarding the

THE RIEMANN HYPOTHESIS

We have seen that Euler solved one of the big challenges of the early 18th century (the *Basel problem*) by proving that

$$1 + (1/2)^2 + (1/3)^2 + (1/4)^2 + (1/5)^2 + \dots = \pi^2/6.$$

He also proved that

$$1 + (1/2)^4 + (1/3)^4 + (1/4)^4 + (1/5)^4 + \dots = \pi^4/90,$$

$$1 + (1/2)^6 + (1/3)^6 + (1/4)^6 + (1/5)^6 + \dots = \pi^6/945,$$

and so on, up to the 26th powers. Generalizing this idea, he defined the *zeta function* $\zeta(k)$ by

$$\zeta(k) = 1 + (1/2)^k + (1/3)^k + (1/4)^k + (1/5)^k + \dots;$$

so $\zeta(2) = \pi^2/6$, $\zeta(4) = \pi^4/90$, $\zeta(6) = \pi^6/945$, etc.
It turns out that $\zeta(k)$ is defined for every real number $k > 1$. However, as we saw earlier, the harmonic series $1 + 1/2 + 1/3 + 1/4 + 1/5 + \dots$ has no finite sum, so $\zeta(1)$ is not defined.

Can we define the zeta function for other numbers, such as 0 or −4 or even the complex number $1/2 + 3i$? In 1859 Riemann found a way of doing so for every real or complex number (except 1), and the function is now known as the *Riemann zeta function.*

It turns out that major problems involving prime numbers are related to the *zeros of the zeta function* – the solutions of the equation $\zeta(z) = 0$ in the complex plane. It also turns out that the zeta function has zeros at −2, −4, −6, −8, … , and that all other zeros lie within a vertical strip between 0 and 1, called the *critical strip.* Moreover, all the *known* zeros in the critical strip (several billion of them, in fact!) occur at points of the form $1/2 + ki$, for some number k, and so they lie on a vertical line known as the *critical line.* So the question arises: *do* all *the zeros in the critical strip lie on this line?* That's the big question that we now call the *Riemann hypothesis.* It is generally believed to be true, but no-one has been able to prove it, even after 150 years.

surrounding higher-dimensional space, he could study manifolds in their own right and measure distances on them. Arising from this work, he was able to describe *infinitely many* different geometries, each one equally valid, and each one a candidate for the physical space we live in.

FUNCTIONS AND SERIES

Other pioneering areas of research arose from Riemann's investigations into which functions can be represented by their Fourier series. This developed the theory of functions of a real variable, raised a problem that inspired Cantor in his famous theory of sets, and led to his definition of the *Riemann integral.*

Part of this investigation was his *rearrangement theorem,* which illustrates the subtlety needed for a study of infinite series. Dirichlet had shown that an infinite series can converge to different answers when we alter the order in which the terms are combined. For example, if we take the series

$$1 - 1/2 + 1/3 - 1/4 + 1/5 - 1/6 + 1/7 - \dots = \log_e 2,$$

and rearrange it so that two positive terms are followed by a negative term, then the resulting series has a different sum:

$$(1 + 1/3) - 1/2 + (1/5 + 1/7) - 1/4 + \dots = 3/2 \log_e 2.$$

Riemann developed this idea by showing that such an infinite series can be rearranged to give any answer whatsoever!

DODGSON

Charles Dodgson (1832–1898) is best known for his children's books *Alice's Adventures in Wonderland* and *Through the Looking-Glass,* written under the pseudonym of Lewis Carroll. He was also an imaginative and pioneering photographer. But his main career was as a mathematics lecturer at Christ Church, one of the colleges of Oxford University, where he wrote extensively on Euclidean geometry, syllogistic logic, the algebra of determinants, and the mathematics of voting.

After growing up in the north of England, Charles Dodgson went to Oxford, where he spent the rest of his life. As a mathematics lecturer from 1856 to 1881, he wrote books and pamphlets on Euclidean geometry, algebra, trigonometry, and other topics, to help his Oxford students in their examinations. He also enjoyed entertaining his friends (both adults and children) with mathematical puzzles, often using these entertainments as a vehicle for conveying serious mathematical ideas.

EUCLIDEAN GEOMETRY

Dodgson was a passionate advocate for Euclid's *Elements,* regarding it as the perfect training for the mind. He wrote commentaries on Books I and II and Book V, and proposed some variations on Euclid's parallel postulate.

In Victorian Britain, the *Elements* was required study for those intending to enter the Church, the Army or the Civil Service, and hundreds of editions were produced. But there was a reaction from those who regarded axiomatic geometry and the learning of proofs as obscure, unsuitable for beginners, and artificial in its insistence on a minimal set of axioms. A growing middle class demanded a more practical approach to mathematics, and the traditional classical education was becoming increasingly irrelevant.

Dodgson entered this debate with a vengeance, producing his most popular geometrical work, *Euclid and his Modern Rivals,* in which he skilfully compared the *Elements,* favourably in each case, with thirteen well-known rival texts. Attempting to reach a wider audience, Dodgson cast his book as a play in four acts.

Not all of Dodgson's geometrical writing was so serious. In *A New Theory of Parallels,* he enthused about the Pythagorean theorem:

> *It is as dazzlingly beautiful now as it was in the day when Pythagoras first discovered it, and celebrated its advent, it is said, by sacrificing a hecatomb of oxen* [100 oxen] — *a method of doing honour to Science that has always seemed to me slightly exaggerated and uncalled-for ... a hecatomb of oxen! It would produce a quite inconvenient supply of beef.*

Charles Dodgson

ALGEBRA

A well-known story, which Dodgson strongly denied, relates how Queen Victoria was so utterly charmed by *Alice's Adventures in Wonderland* that she demanded:

Send me the next book Mr Carroll produces.

The next book duly arrived: it was entitled

An Elementary Treatise on Determinants with their Application to Simultaneous Linear Equations and Algebraical Geometry.

Queen Victoria was not amused.

As its title suggests, determinants can be used in the solution of simultaneous equations. If a, b, c and d are numbers, then their *determinant* $\begin{vmatrix} a & b \\ c & d \end{vmatrix}$ is the number $ad - bc$, and there are analogues for larger arrays of numbers. Dodgson invented a useful method, which he called the *condensation method*, to transform the determinants of such larger arrays into several smaller ones of the type shown above; it is still used today.

VOTING

In the 1870s Dodgson became greatly involved in the theory of voting and elections. A strong supporter of proportional representation, at a time when most Parliamentary constituencies (including Oxford University) were represented by two or more members, he wrote a pamphlet explaining why several widely used voting systems, such as the simple majority and the single transferable vote, can fail to give fair results.

Many years later the Oxford philosopher Michael Dummett regretted that Dodgson had never completed the book that he planned to write on the subject:

Such were the lucidity of exposition and his mastery of the topic that it seems possible that, had he ever published it, the political theory of Britain would have been significantly different.

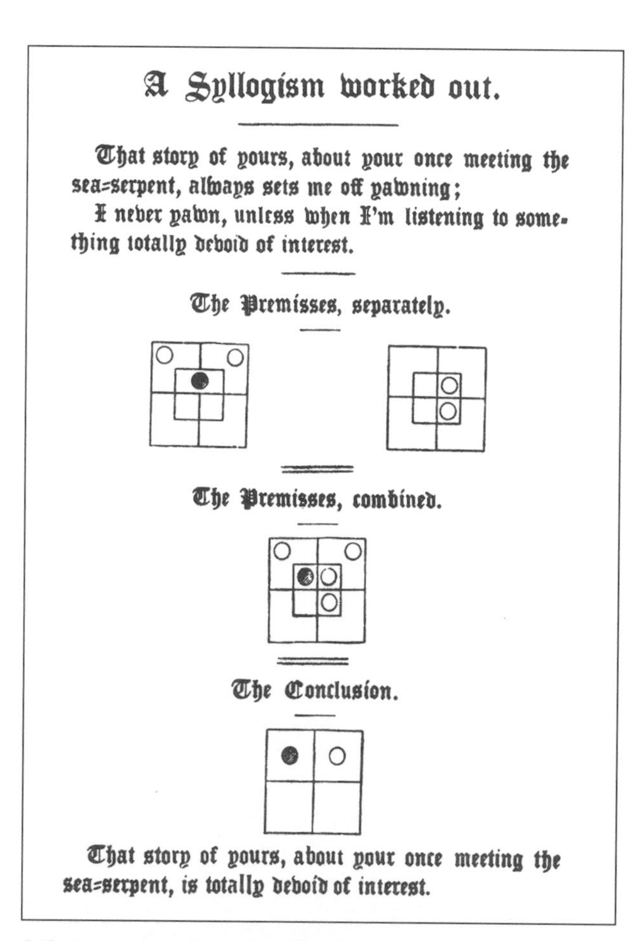

A Syllogism worked out.

That story of yours, about your once meeting the sea-serpent, always sets me off yawning;

I never yawn, unless when I'm listening to something totally devoid of interest.

The Premisses, separately.

The Premisses, combined.

The Conclusion.

That story of yours, about your once meeting the sea-serpent, is totally devoid of interest.

Solving a complicated syllogism

LOGIC

In his later years Dodgson spent much of his time presenting symbolic logic as an entertainment for children to develop their powers of logical thought, and as a serious topic of study for adults.

Much of his work was concerned with Aristotelian-style syllogisms, with a pair of premises that lead to a conclusion. Dodgson's designed his syllogisms to be entertaining — for example, from the two premises

A prudent man shuns hyenas.

No banker is imprudent

he deduced the conclusion

No banker fails to shun hyenas.

Dodgson developed a method for solving such syllogisms, using a board with counters that were placed according to certain rules, and he used these to instruct both adults and children. He then extended his ideas to situations where there are many more premises — in one example, up to fifty — and one is required to find the conclusion.

CANTOR

Georg Cantor

The creation of modern set theory is due to Georg Cantor (1845–1918). He established the importance of one-to-one correspondences between sets and founded the theory of transfinite numbers, showing, in particular, that infinities can have different sizes. This work arose from his various investigations into the convergence of Fourier series and into whether just one trigonometrical series can represent a given function.

Cantor was born in St Petersburg and began his university studies at the Polytechnic in Zürich, moving after a year to the more prestigious Berlin University where he took his doctorate. In 1869 he became a lecturer at the University of Halle, being promoted to professor ten years later. Although he had always hoped to obtain a position in Berlin, he remained in Halle for the rest of his life, teaching there for many years before succumbing to severe mental illness.

SET THEORY

Cantor introduced his theory of sets in a number of papers dating from 1874. For him, a set was

> *any collection into a whole M of definite and separate objects m of our intuition or of our thought.*

The objects *m* making up the set are called its *elements*. With this very abstract definition, a set could have as elements many different kinds of things, such as the people in the world, the positive integers, or the real numbers. A *subset* of a set *B* is a set whose elements are also elements of *B*.

Two sets *A* and *B* are said to be *equivalent* – we can think of this as meaning that they have the same size – if we can match them up exactly; that is, there is a one-to-one correspondence between the elements of *A* and those of *B*. If the sets *A* and *B* are finite then they must have the same number of elements, but if the sets are infinite then things become much more interesting! In particular, we see opposite that the set of integers and the set of fractions are equivalent, whereas the set of integers (or fractions) and the set of real numbers are not equivalent – they have different *cardinalities*. As we see later, the *Continuum hypothesis* is the conjecture that every infinite subset of the real numbers is equivalent to either the set of integers (or fractions) or the set of real numbers.

Although Cantor's work on infinite sets caused great controversy at first, it was soon taken up by other mathematicians and was found to have important applications throughout the whole of mathematics.

SOME INFINITIES ARE LARGER THAN OTHERS

In his *Two New Sciences,* Galileo noted that *the set of positive integers* (1, 2, 3, ...) *is larger than the set of their squares* (1, 4, 9, ...), and yet *the two sets must have the same size* since we can match them up exactly: 1 ↔ 1, 2 ↔ 4, 3 ↔ 9, 4 ↔ 16, ...

We can also match up the positive integers with some larger sets, such as *all* the integers (positive, negative and zero), by listing them in the order: 0, 1, –1, 2, –2, 3, –3, 4, ... ; notice that every integer occurs somewhere in this list. A set that can be matched up with the set of positive integers in this way is called *countable,* because we can list (or count) all its elements – so *the set of all integers is countable.*

Now look at all the fractions. Although this set seems much larger than the set of positive integers, Cantor made the unexpected discovery that we can list all the fractions in order – so *the set of all fractions is countable.* On the other hand, as he also proved, *the set of all real numbers is not countable.* It follows that the set of real numbers is strictly larger than the set of all fractions – and so *some infinite sets are bigger than others.* Cantor then took this idea further by proving that *there are infinitely many infinities, all of different sizes.*

THE SET OF ALL FRACTIONS IS COUNTABLE

We first list all the *positive* fractions, as shown: the first row lists the integers, the second row lists the 'halves', and so on. We then 'snake around' the diagonals of this array of numbers, deleting any numbers that we have seen before: this gives the list

$$1, 2, \tfrac{1}{2}, \tfrac{1}{3}, 3, 4, 1\tfrac{1}{2}, \tfrac{2}{3}, \tfrac{1}{4}, \tfrac{1}{5}, 5, 6, 2\tfrac{1}{2}, \ldots$$

This list contains all the positive fractions.

To list *all* the fractions (positive, negative and zero) in order, we then alternate + and −, as before. So the set of all fractions is countable.

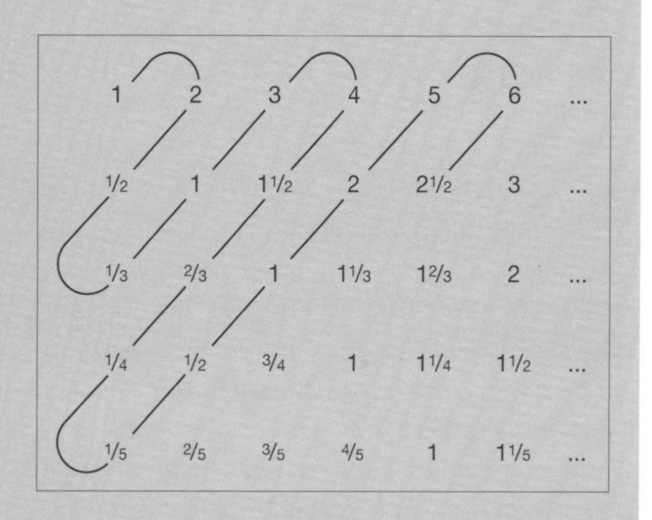

THE SET OF ALL REAL NUMBERS IS NOT COUNTABLE

It is enough to prove that *the set of all numbers between* 0 *and* 1 *is not countable.*

To do so, we assume that this set is countable, and obtain a contradiction.

So, assuming that this set is countable, we can list its numbers (as decimals), as follows.

$$0{\cdot}a_1\, a_2\, a_3\, a_4\, a_5 \ldots,\ 0{\cdot}b_1\, b_2\, b_3\, b_4\, b_5 \ldots,\ 0{\cdot}c_1\, c_2\, c_3\, c_4\, c_5 \ldots,\ 0{\cdot}d_1\, d_2\, d_3\, d_4\, d_5 \ldots, \text{ etc.}$$

By our assumption, this list contains all the numbers between 0 and 1.

We obtain the required contradiction by constructing a new number between 0 and 1 that does not lie in this list. To do this, we choose numbers $X_1, X_2, X_3, X_4, \ldots$ from 1 to 9 such that $X_1 \neq a_1$, $X_2 \neq b_2$, $X_3 \neq c_3$, $X_4 \neq d_4, \ldots$, and consider the number $0{\cdot}X_1\, X_2\, X_3\, X_4 \ldots$.

Since $X_1 \neq a_1$, this new number differs from the first number in the list;

since $X_2 \neq b_2$, it differs from the second number in the list; and so on.

Thus, this new number differs from every number in the list. This gives us the required contradiction.

So the set of all numbers is not countable.

KOVALEVSKAYA

The mathematician and novelist Sonya Kovalevskaya (1850–1891) made valuable contributions to mathematical analysis and partial differential equations. With no higher education available for women in Russia, she went to Heidelberg, attending lectures of Kirchhoff and Helmholtz, and then to Berlin, where she worked with Weierstrass, later becoming the first female professor in Stockholm. She won the coveted Prix Bordin of the French Academy for a memoir on the rotation of bodies.

Sonya (or Sofya) Krukovskaya was born into a noble family, the daughter of an artillery general. At Palabino, the family's large country estate where she grew up, there was insufficient wallpaper available to cover the nursery wall, and the task was completed with some old calculus notes of her father's:

> *These sheets, spotted over with strange incomprehensible formulas soon attracted my attention. I remember how, in my childhood, I passed whole hours before that mysterious wall, trying to decipher even a single phrase, and to discover the order in which the sheets ought to follow each other.*

At that time Russian universities were not open to women. The only way for Sonya Krukovskaya to continue her studies was to travel abroad, and in order to obtain parental permission for this, she arranged a 'marriage of convenience' with a young palaeontologist named Vladimir Kovalevsky.

GERMANY

The newly-weds travelled first to Heidelberg, where Kovalevskaya unofficially attended the physics lectures of Kirchhoff and Helmholtz and the mathematics ones of Leo Königsberger, a former student of Karl Weierstrass in Berlin.

Two years later, in 1871, her husband went to Jena, while she moved to Berlin to study mathematics with Weierstrass. Again she discovered that she was barred from attending lectures, but after she presented Königsberger's glowing letter of recommendation to Weierstrass, he became greatly impressed with her mathematical powers. He agreed to take her on as a private student, providing her with lecture notes for the lectures she was missing, and working with her on a range of mathematical topics.

Sonya Kovalevskaya

Their collaboration lasted for four years, during which Kovalevskaya wrote three outstanding research papers. One was ground-breaking work on the solutions of partial differential equations, and contained a result now known as the *Cauchy–Kovalevskaya theorem.* The second, in which she generalized some results of Euler, Lagrange and Poisson, involved a type of integral called an Abelian integral. The third, on Saturn's rings, extended earlier work of Laplace.

These three brilliant papers qualified her for a doctorate from the University of Göttingen, and because of the exceptional nature of the first paper and her claimed lack of proficiency in German, her oral examination was waived. But in spite of enthusiastic references from Weierstrass, Kovalevskaya was unable to find an academic post in central Europe and returned to Russia.

STOCKHOLM

Matters were not well between Sonya and her husband. They had a daughter, but then separated. Unable to secure regular employment, Vladimir became involved with some shady financial deals and eventually went bankrupt and committed suicide.

In desperation, Kovalevskaya turned to Weierstrass for help, and through the good offices of one of his former students, Gösta Mittag-Leffler, she was awarded a position at Stockholm University in Sweden. Although some (such as the writer August Strindberg) were bitterly opposed to the appointment of a female academic, the local newspaper was enthusiastic:

> *Today we do not herald the arrival of some vulgar insignificant prince of noble blood. No, the Princess of Science, Madam Kovalevskaya, has honoured our city with her arrival. She is to be the first woman lecturer in all Sweden.*

Karl Weierstrass

From this moment, her situation began to improve. The courses she taught on Weierstrass's analysis were enjoyed by her students, and she became increasingly involved with new researches on the refraction of light. She also returned to writing novels, an activity she had pursued in her youth.

The high point of her career occurred in 1888. The French Academy of Sciences had announced the topic for its prestigious Prix Bordin, and her submission, *On the rotation of a solid body about a fixed point,* involving the solution of complicated systems of differential equations, won the prize. So outstanding was her submission that the value of the prize was increased from 3000 to 5000 francs.

But eventually things started to change. She began to find Stockholm too narrow and provincial and yearned to go to Paris, a city that she loved. The Swedish climate, with its cold winters and overlong summer evenings, was uncongenial. She became depressed and failed to take proper care of herself. Eventually, she succumbed to influenza and pneumonia and died at the early age of 41.

KLEIN

Felix Klein (1849–1925) was a German mathematician who worked on geometry – particularly non-Euclidean geometry and the connection between geometry and group theory. At the University of Göttingen he developed the world's foremost mathematical centre, where he was an influential educationalist and teacher. He was founder of the great mathematical *Encyklopädie* and was editor of one of the leading mathematical journals of the time, the *Mathematische Annalen.*

Klein was born in Düsseldorf and studied at Bonn, Göttingen and Berlin. From 1872 to 1875 he was professor at Erlangen before moving to Munich, Leipzig, and finally to the University of Göttingen in 1886. As head of the mathematics department there, Klein proved to be a worthy successor to Gauss, Dirichlet and Riemann. His school of mathematics was the most famous in the world and attracted many brilliant scholars.

Göttingen began admitting women in 1893, and one of Klein's doctoral students was Grace Chisholm (later, Grace Chisholm Young after her marriage), who gave an insight into the views held there:

> *Professor Klein's attitude is this, he will not countenance the admission of any woman who has not already done good work, and can bring proof of the same in the form of degrees or their equivalent ... and further he will not take any further steps till he has assured himself by a personal interview of the solidity of her claims. Professor Klein's view is moderate. There are members of the Faculty here who are more eagerly in favour of the admission of women and others who disapprove altogether.*

Klein dominated the institutional developments in German mathematics and was a prime mover for the organization of mathematics conferences. He also led an international team in the production of the multi-volume *Encyclopedia of Mathematical Science*, published between 1890 and 1920.

THE *ERLANGER PROGRAMM*

By 1870 the world of geometry had become very complicated. In addition to non-Euclidean geometry, there were Euclidean and spherical geometries, similarity, affine and projective geometries, and many others besides. The remaining years of the century saw a variety of attempts to sort out the confusion and impose order on the subject.

The most famous of these was the *Erlanger Programm,* circulated in written form at Klein's inaugural lecture as professor at the University of

The Mathematics Club of Göttingen in 1902: at the table are David Hilbert, Felix Klein, Karl Schwarzschild and Grace Chisholm

THE KLEIN BOTTLE

Klein's most familiar legacy is the surface known as a *Klein bottle*. It is constructed from a Möbius band by gluing its boundary edge to the boundary of a circular disc, and cannot exist in three dimensions without intersecting itself.

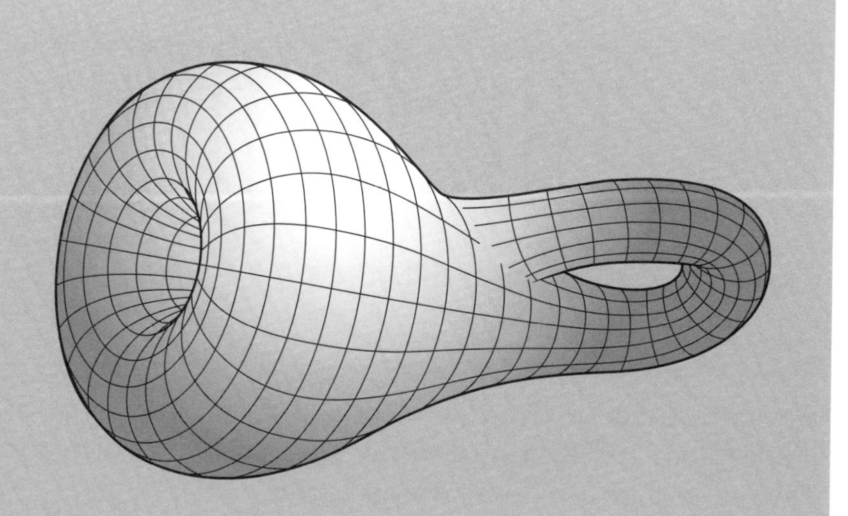

Erlangen in 1872, when he was 23. The lecture was aimed at a university-wide audience and was concerned with Klein's pedagogical views on mathematical education, the unity of all knowledge, and the importance of a complete and wide-ranging education.

Felix Klein

In the *Erlanger Programm* Klein gave a new and remarkable definition of a 'geometry' that helped to unify existing geometries and also provided a 'road map' for future research. For him, a *geometry* was a set of points (such as the points in a plane) with transformations (such as rotations, reflections and translations) defined upon it, where we are interested in those properties of the set that remain unchanged by the transformations:

> *Given a manifold* [the set of points] *and a group of transformations of the same; to investigate the configurations belonging to the manifold with regard to such properties as are not altered by the transformations of the group.*

For our example in the plane we obtain our familiar Euclidean geometry. Since our transformations do not alter size or shape, we are interested in such geometrical properties as the lengths of lines and the congruence between triangles.

If we now enlarge the group of transformations we can obtain other geometries. For example, if we also include *scalings* (that make figures larger or smaller), then we obtain a different geometry, sometimes called *similarity geometry.* Here the transformations preserve shape (but not size, in general), so fewer geometrical properties are preserved.

Continuing to enlarge the group of transformations produces a 'hierarchy of geometries', all of which turn out to be contained in projective geometry. In particular, Euclidean and non-Euclidean geometries are both special cases of projective geometry, so any result in projective geometry is also true for them (and indeed for all other geometries). The task that Klein set himself, of unifying all geometries, had been achieved.

In 1923 Klein summarized his lifelong attitude to geometry:

> *I did not conceive of the word geometry one-sidedly as the subject of objects in space, but rather as a way of thinking that can be applied with profit in all domains of mathematics.*

CHAPTER 5

THE MODERN AGE

In our final chapter we meet the mathematicians who have:

- **examined the limits of what we can prove and shown why some tasks are impossible to carry out**
- **laid the foundations of our current scientific knowledge**
- **carried out mathematical work of historical, social and political impact, and thereby changed the world we live in**
- **developed computers, both theoretical and practical, that enable us to simulate, model and prove things that we could not do otherwise, while raising questions about our identity.**

PARADOXES AND PROBLEMS

In previous chapters we have seen a developing desire to place mathematics on a sounder foundation, with the story going from the underpinning notions of the calculus to arithmetic and the theory of sets. As 20th-century mathematicians examined more carefully the nature of infinity and the problems connected with sets they met a number of problems and paradoxes. One of the most famous of these was

The Princeton Institute for Advanced Study, established in 1930: 25 Nobel Laureates and 38 (out of 52) Fields Medallists have been affiliated with it

G. H. Hardy with his research student Mary Cartwright, later to be the first woman appointed President of the London Mathematical Society

formulated by Bertrand Russell in 1902, and necessitated a much more thorough treatment of the very foundations of set theory and of the exact nature of deductive proof.

Another approach was taken by David Hilbert, whose attempt to make arithmetic secure was to make it *axiomatic,* an approach that he had already used with success when dealing with the foundations of geometry. Instead of defining all the basic terms, such as point or line, he gave a set of rules (or axioms) that they had to satisfy.

Although Hilbert's approach was influential, his objectives were eventually proved to be unattainable, as demonstrated in the 1930s by Kurt Gödel and Alan Turing, who obtained a number of amazing and unexpected results about the limits of what can be proved or decided.

ABSTRACTION AND GENERALIZATION

The 19th-century trend towards increasing generalization and abstraction continued to accelerate dramatically throughout the 20th century. For example, Albert Einstein used the abstract formulations of geometry and calculus for his general theory of relativity, while algebra became an abstract and axiomatic subject, being particularly influenced by the work of Emmy Noether. Advances also continued to be made in number theory, with Hardy (and his co-workers Littlewood and Ramanujan) and Andrew Wiles making major contributions.

Meanwhile, new areas of the subject came into being, such as algebraic topology and the theory of 'Hilbert spaces', while machine computation entered the mainstream of the subject, as spectacularly illustrated by Appel and Haken in their proof of the four-colour theorem.

SPREAD AND DEVELOPMENT

The 20th century saw mathematics becoming a major profession throughout the world, with jobs in education and industry and numerous areas of specialization and application.

With mathematics developing at such a fast rate, many new journals have been created, and national and international conferences have become widespread. Most important among these meetings are the International Congresses of Mathematicians, held every four years, when the prestigious Fields medals are awarded and many thousands of mathematicians gather to learn about the latest developments in their subject.

HILBERT

On 8 August 1900, David Hilbert (1862–1943), one of the greatest mathematicians of the day, gave the most celebrated mathematical lecture of all time. For it was on this date, at the International Congress of Mathematicians in Paris, that he presented a list of unsolved problems for 20th-century mathematicians to tackle. Trying to solve these problems helped to set the mathematical agenda for the next hundred years.

David Hilbert was born in Königsberg in Eastern Prussia and received his doctorate there in 1885. After teaching in Königsberg for a few years, he was invited by Felix Klein to join the faculty at Göttingen, where he spent the rest of his life.

His mathematical range was immense — from abstract number theory and invariant theory, via the calculus of variations and the study of analysis (and so-called 'Hilbert spaces'), to potential theory and the kinetic theory of gases.

THE FOUNDATIONS OF GEOMETRY

Following Cantor's introduction of set theory and subsequent investigations by various mathematicians into the foundations of arithmetic, Hilbert became increasingly involved with the foundations of geometry.

Although Euclid's axiom system had worked well for two thousand years, it contained a number of unwarranted assumptions. Hilbert duly set about replacing it by alternative sets of axioms that were completely foolproof. His aim, in particular, was to find axiom systems that are

- *consistent:* the axioms do not lead to contradictions
- *independent:* no axiom can be deduced from the others
- *complete:* any statement that we may formulate within the system can be proved to be either true or false

In 1899 Hilbert produced his influential *Grundlagen der Geometrie* (Foundations of Geometry), in which he developed his axiom systems for Euclidean and projective geometry. Four years later he produced a second edition in which he also axiomatized non-Euclidean geometry.

Hilbert had a grand plan. He was convinced that the whole of classical mathematics could be similarly axiomatized, and with Paul Bernays he wrote a two-volume work with this purpose in mind. But as they progressed, they experienced unexpected difficulties with the details of their arguments, and it soon became apparent that Hilbert's plan was doomed to failure.

THE HILBERT PROBLEMS

Who of us would not be glad to lift the veil behind which the future lies hidden: to cast a glance at the next advances of our science and at the secrets of its development during future centuries?

So asked David Hilbert in his famous address at the Paris Congress, at which he presented his list of twenty-three unsolved problems. We have already met one of these problems, the Riemann hypothesis, which remains unsolved to this day. Here we present a few more, some of which will be discussed later in this chapter.

Problem 1: *Prove the Continuum hypothesis, that there is no set whose cardinality lies strictly between those of the integers and the real numbers.*

We recall that Cantor proved that infinities can have different sizes, and that the set of real numbers is strictly larger than the set of integers (or fractions). This problem asks us to prove that no infinite set is larger than the set of integers but smaller than the set of real numbers.

Problem 2: *Are the axioms of arithmetic consistent?*

Hilbert based his treatment of the consistency of his geometrical axioms on the assumption that arithmetic (that is, our real number system) can be similarly axiomatized. This problem asks whether this latter assumption is valid, or whether there could be, 'somewhere out there', a contradiction that we never expected.

Problem 3: *Given two polyhedra with the same volume, can we always cut the first into finitely many pieces that can then be reassembled to give the second?*

In 1833, János Bolyai proved that if two polygons have the same area, then the first can be cut into pieces that can be rearranged to give the second; the following example shows a triangle reassembled to give a square. This problem asks whether a similar result holds in three dimensions.

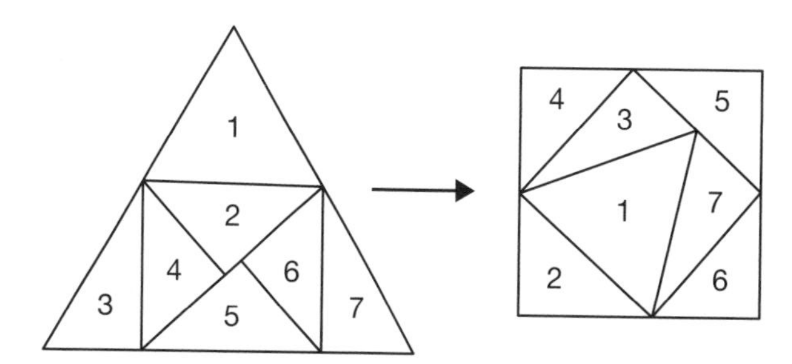

The answer is no. Within two years Max Dehn proved that a regular tetrahedron *cannot* be cut into pieces that can then be reassembled to give a cube with the same volume.

Problem 18: *What is the most efficient way to stack spheres so that the amount of empty space between them is as small as possible?*

This problem was considered by Harriot and Kepler. Two ways to stack the spheres are cubic stacking and hexagonal stacking, but neither is the most efficient. It turns out that the way your greengrocer stacks oranges is the most efficient —the proportion of empty space is about 0.36, which is less than the 0.48 and 0.40 proportions of the other two. But to prove this rigorously was horrendous: in 1998 Thomas Hales gave a computer-aided proof that involved three gigabytes of computer power.

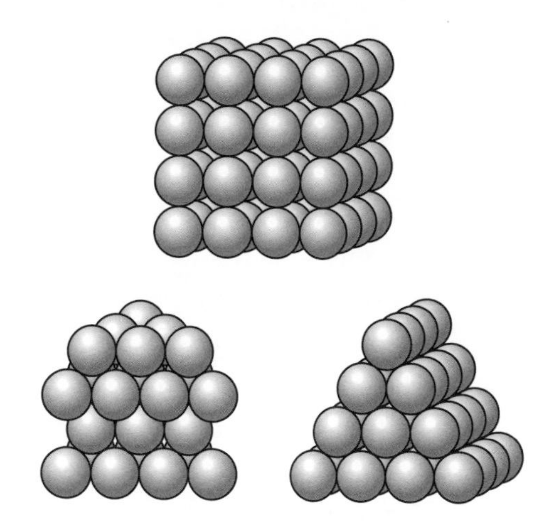

Cubic stacking, hexagonal stacking and greengrocer's stacking

POINCARÉ

Henri Poincaré (1854–1912) is viewed as one of the great geniuses of all time, being probably the last person to cover the entire range of mathematics. He virtually founded the theories of several complex variables and algebraic topology, and one of his conjectures in topology, known as the *Poincaré conjecture*, was solved only in this century. He made outstanding contributions to differential equations and non-Euclidean geometry, and also worked on electricity, magnetism, quantum theory, hydrodynamics, elasticity, the special theory of relativity and the philosophy of science. As an active popularizer of his subject, he wrote popular works for non-mathematicians, stressing the importance of mathematics and science and discussing the psychology of mathematical discovery.

Poincaré was born in Nancy, Northern France, and displayed great ability and interest in mathematics from a young age. He came from a distinguished family, and his cousin, Raymond Poincaré, became President of the French Republic during the First World War. He attended the École Polytechnique in 1873, and after graduating went to the École des Mines for further study. In 1879 he obtained a post at the University of Caen, moving two years later to the University of Paris, where he remained until his death at the age of 58.

Henri Poincaré

KING OSCAR'S PRIZE

Oscar II, King of Sweden and Norway, was an enthusiastic patron of mathematics. To mark his 60th birthday, he offered a prize of 2500 Swedish crowns for a memoir on any of four given topics, one of which was on predicting the future motion of a system of bodies moving under mutual gravitational attraction:

ANALYSIS SITUS

Although topology had its origins in the problem of the bridges of Königsberg and the theory of polyhedra, it was Poincaré and his successors who transformed it into a rich and powerful new way of looking at geometrical objects. Poincaré's work *Analysis Situs* (Analysis of Position), published in 1895, used algebraic methods to distinguish between surfaces and was an early systematic account of what we now call *algebraic topology*.

Given a system of arbitrarily many mass points that attract each according to Newton's law, under the assumption that no two points ever collide, try to find a representation of the coordinates of each point as a series in a variable that is some known function of time and for all of whose values the series converges uniformly.

Newton had solved this problem for two bodies, and Poincaré responded to the King's challenge by attacking a special case of the problem when there are only three bodies (the *restricted three-body problem*), hoping that he would eventually be able to generalize his results to the general three-body problem, and then to more than three bodies.

By considering approximations to the orbits, he was able to make considerable progress, developing valuable new techniques in analysis along the way. Although he could not solve the three-body problem in its entirety, he developed so much new mathematics in his attempts that he was awarded the prize.

However, while his paper was being prepared for publication, one of the editors queried it, unable to follow Poincaré's arguments. Poincaré realized that he had made a mistake: contrary to what he formerly thought, even a small change in the initial conditions can produce vastly different orbits. This meant that his approximations did not give him the results he had expected. But this led to something even more important. The orbits that Poincaré discovered were what we now call *chaotic:* he had stumbled on the mathematics at the basis of modern-day *chaos theory,* where even with deterministic laws the resulting motion may be irregular and unpredictable.

THE POINCARÉ DISC

One difficulty caused by the non-Euclidean geometry of Bolyai and Lobachevsky was that it was hard to visualize. A number of pictorial representations were suggested, the most successful being the 'disc model', discovered by Poincaré in 1880.

Consider the following picture of a disc (the inside of a circle). We consider a geometry in which the *points* are those lying inside the bounding circle, and the *lines* are either diameters that pass through the centre of the disc or circular arcs that meet the bounding circle at right angles.

As we can see,

- some pairs of *lines* (diameters or circular arcs) do not meet
- some pairs of *lines* meet at internal *points*
- some pairs of *lines* meet on the bounding circle: such *lines* are said to be *parallel.*

With these definitions of *point* and *line,* Euclid's first four axioms are satisfied, but not the fifth, so we do indeed have a non-Euclidean geometry. In this geometry many Euclidean concepts (such as size and shape) are no longer appropriate; for example, all the grey and white triangles in the picture turn out to be congruent to each other! The Dutch artist Maurits Escher based some of his woodcuts (such as *Circle Limit IV*) on this picture of a non-Euclidean geometry.

RUSSELL AND GÖDEL

Bertrand Russell (1872–1970) and Kurt Gödel (1906–1978) were the most important logicians of the 20th century. In the early years of the century the foundations of mathematics were in turmoil. Russell's celebrated paradox caused major difficulties which took time to resolve. Meanwhile, as Hilbert and others were continuing with their ambitious programme of sorting out the foundations of arithmetic, Gödel upset the apple-cart with his startling results on the completeness and consistency of axiom systems.

Bertrand Russell was one of the outstanding figures of the 20th century. Born of a noble family, he was orphaned as a young child and brought up by his grandmother, before going to Trinity College, Cambridge, to study both mathematics and the moral sciences. A vigorous peace campaigner, he was twice imprisoned for his anti-war activities. In 1950 he was awarded the Nobel Prize for Literature.

Kurt Gödel was born in Vienna. He suffered from rheumatic fever at the age of 6, and from then on was constantly obsessed with his health. After graduating from the University of Vienna in mathematics, he wrote his doctoral thesis there on mathematical logic and joined the faculty. In 1940 he emigrated to the USA, where he spent the rest of his distinguished career at the Princeton Institute for Advanced Study, receiving many awards and becoming a close friend of Albert Einstein. He suffered from paranoia and, convinced that he was being poisoned, refused to eat and died of malnutrition.

Bertrand Russell as a young man

Kurt Gödel receives an award from Albert Einstein

RUSSELL'S PARADOX

The German logician Gottlob Frege was on the point of sending a book on axiomatics to his publishers when he received a letter from Bertrand Russell, effectively saying:

Dear Gottlob, Consider the set of all sets that are not members of themselves. Bertrand.

This paradox demolished much of his book, and mathematical logic was changed for ever.

A simpler version of Russell's paradox concerns a village barber who shaves all those who do not shave themselves, but does not shave those who do shave themselves. The question arises: *Who shaves the barber?*

- If the barber shaves himself, we obtain a contradiction since he does not shave those who shave themselves
- But if the barber does not shave himself, we obtain a contradiction since he shaves all those who don't shave themselves

Thus, this problem can have no answer.

Russell's version of his paradox had asked: *let S be the set of all sets that are not members of themselves. Is S a member of itself?* Whether the answer is *yes* or *no*, we obtain contradictions similar to those above.

KURT GÖDEL AND HILBERT'S PROBLEMS

In 1931 Gödel produced a paper that changed mathematics for ever. His first 'bombshell', the *Incompleteness theorem,* was to prove that:

> ***In any axiomatic system that includes the integers, there are true results that cannot be proved, and there are 'undecidable' results that cannot be proved either true or false.***

Hilbert's Problem 1

We saw earlier that Hilbert asked for a proof of the *Continuum hypothesis,* that no set is larger than the set of integers but smaller than the set of real numbers. Using his Incompleteness theorem, Gödel proved that, if we use Zermelo–Fraenkel set theory, then *the Continuum hypothesis cannot be disproved.* But in 1963 the American mathematician Paul Cohen stunned the mathematical world (thereby winning a Fields Medal) by proving that, under the same conditions, *the Continuum hypothesis cannot be proved.* Combining these results, we deduce that:

> ***The Continuum hypothesis cannot be proved either true or false – it is 'undecidable'.***

Hilbert's Problem 2

We recall that this asked us to prove that the axioms of arithmetic are consistent, so that contradictions cannot occur — but then Gödel produced a second bombshell by proving that:

> ***The consistency of any theory that includes the integers cannot be proved within the theory itself* – in other words, *we cannot prove that contradictions can never occur.***

One might have thought that these results would have finished off the subject for good – but most mathematicians chose to ignore them and carried on regardless.

The appearance of Russell's paradox led philosophers and mathematicians to study set theory more carefully, and a number of versions emerged that could deal with such paradoxes more or less satisfactorily. Of these, the most successful and universally accepted was the *Zermelo–Frankel set theory,* due originally to Ernst Zermelo of Göttingen, and revised by Adolf Fraenkel of Marburg.

86 CARDINAL ARITHMETIC [PART III

***110·632.** $\vdash : \mu \in \mathrm{NC} \,.\supset.\, \mu +_c 1 = \hat{\xi}\{(\exists y) \,.\, y \in \xi \,.\, \xi - \iota\text{‘}y \in \mathrm{sm}\text{“}\mu\}$

Dem.

$\vdash . *110{\cdot}631 . *51{\cdot}211{\cdot}22 . \supset$

$\vdash : \mathrm{Hp} \,.\supset.\, \mu +_c 1 = \hat{\xi}\{(\exists \gamma, y) \,.\, \gamma \in \mathrm{sm}\text{“}\mu \,.\, y \in \xi \,.\, \gamma = \xi - \iota\text{‘}y\}$

[*13·195] $= \hat{\xi}\{(\exists y) \,.\, y \in \xi \,.\, \xi - \iota\text{‘}y \in \mathrm{sm}\text{“}\mu\} : \supset \vdash . \mathrm{Prop}$

***110·64.** $\vdash . 0 +_c 0 = 0$ [*110·62]

***110·641.** $\vdash . 1 +_c 0 = 0 +_c 1 = 1$ [*110·51·61 . *101·2]

***110·642.** $\vdash . 2 +_c 0 = 0 +_c 2 = 2$ [*110·51·61 . *101·31]

***110·643.** $\vdash . 1 +_c 1 = 2$

Dem.

$\vdash . *110{\cdot}632 . *101{\cdot}21{\cdot}28 . \supset$

$\vdash . 1 +_c 1 = \hat{\xi}\{(\exists y) \,.\, y \in \xi \,.\, \xi - \iota\text{‘}y \in 1\}$

[*54·3] $= 2 . \supset \vdash . \mathrm{Prop}$

The above proposition is occasionally useful. It is used at least three times, in *113·66 and *120·123·472.

PRINCIPIA MATHEMATICA

Between 1910 and 1913 Bertrand Russell and his Cambridge colleague Alfred North Whitehead wrote a pioneering three-volume work entitled *Principia Mathematica.* Based partly on ideas of Cantor and Frege, it was designed to deduce the whole of mathematics from a small number of basic principles. Above is their proof of the proposition '1 + 1 = 2'.

EINSTEIN AND MINKOWSKI

Albert Einstein (1879–1955), an iconic figure of the 20th century, was the greatest mathematical physicist since Isaac Newton. He revolutionized physics with his theories of special and general relativity. These drew on mathematical ideas, not previously used in physics, some of which had been developed by Riemann and by Hermann Minkowski (1864–1909).

Albert Einstein — a plaque in Ulm

Einstein was born in Ulm, Southern Germany moving to Munich the next year. He was slow in learning to speak and showed little promise in his early schooling. He was admitted to Zürich polytechnic at his second attempt in 1896 to a course for mathematics and science teachers and graduated in 1900. Although one of his lecturers was Minkowski, he gained little from the formal teaching and preferred to read independently and think deeply about the fundamental ideas and assumptions of physics. After graduation he supported himself by part-time teaching until he obtained a position in the Swiss Patent Office in Bern.

In 1905 Einstein submitted his paper on special relativity to the University of Bern in support of his application for a doctorate, and it was rejected! However recognition of his work soon arrived as it became more widely known. He then held positions at the Universities of Zürich, Prague and Berlin, and announced his general theory of relativity in 1915. He was awarded the Nobel Prize in 1921 for his work on quantum theory, rather than relativity. In 1933 he went to America and from then on was based at the Institute for Advanced Study in Princeton.

EINSTEIN'S *ANNUS MIRABILIS*

In 1905, his 'year of wonders', Albert Einstein published four papers of ground-breaking importance. First he published the work that introduced quanta of energy — that light can be absorbed or emitted only in discrete amounts, a core idea of quantum theory. Next was a paper on Brownian motion, explaining the movement of small particles suspended in a stationary liquid.

His third paper, on the electrodynamics of moving bodies, introduced a new theory linking time, distance, mass and energy. It was consistent with electromagnetism, but omitted the force of gravity. This became known as the *special theory of relativity* and assumed that c, the speed of light, is constant, irrespective of where you are or how you move.

On 21 November 1905 he published *Does the Inertia of a Body Depend Upon Its Energy Content?* This contains one of the most famous equations of all, $E = mc^2$, asserting the equivalence of mass and energy.

MINKOWSKI AND SPECIAL RELATIVITY

Minkowski was born of German parents in Lithuania. In 1902 he moved to the University of Göttingen, where he became a colleague of Hilbert. He developed a new view of space and time and laid the mathematical foundations of the theory of relativity. Minkowski described his approach as follows:

> *Henceforth space by itself, and time by itself, are doomed to fade away into mere shadows, and only a kind of union of the two will preserve an independent reality.*

The *kind of union* that Minkowski mentions is now known as *space-time* and is a four dimensional non-Euclidean geometry that incorporates the three dimensions of space with the one of time. It comes with a way of measuring the distance between two different points of space-time. Space and time are now no longer separate, as Newton had thought, but are intermixed. A reviewer said of his work that

> *purely mathematical considerations, including harmony and elegance of ideas, should dominate in embracing new physical facts. Mathematics, so to speak, was to be master and physical theory could be made to bow to the master.*

Hermann Minkowski

Below is a simplified diagram of space-time with only one space dimension going horizontally and with time going vertically. In Euclidean geometry the distance of each point (x, t) to the origin is $\sqrt{(x^2 + t^2)}$, but the requirements of relativity replace this in space-time with the distance $\sqrt{(x^2 - c^2t^2)}$. The minus sign implies that events in space-time, such as the one labelled 'here and now', are associated with two cones. With just one space dimension, these cones are now triangles, with one representing the future of the 'here and now' and the other its past.

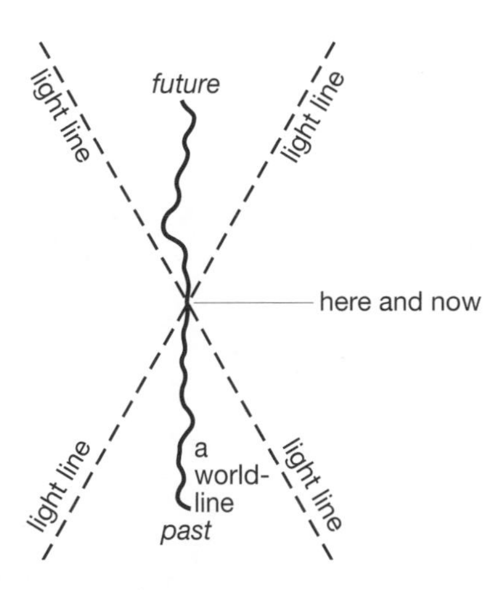

GENERAL RELATIVITY

Einstein initially thought little of Minkowski's approach to space-time, but later found it invaluable, indeed essential, when he was trying to extend his theory to include gravity. His *general theory of gravity,* building also on Riemann's geometrical ideas, produced space-time that was curved as a result of the presence of mass and energy. The curvature increased near to massive bodies, and it was the curvature of space-time that controlled the motion of bodies.

The theory predicted that light rays would be bent by the curvature of space-time produced by the sun, an effect that was observed during the 1919 eclipse of the sun.

HARDY, LITTLEWOOD AND RAMANUJAN

The lengthy and fruitful collaboration of G. H. Hardy (1877–1947) and J. E. Littlewood (1885–1977) was the most productive in mathematical history. Dominating the English mathematical scene for the first half of the 20th century, they produced a hundred joint papers of great influence, most notably in analysis and number theory. Into their world came Srinivasa Ramanujan (1887–1920), one of the most brilliant and intuitive mathematicians of all time, who left India to work with them in Cambridge until his untimely death at the age of 32.

Hardy and Littlewood at Trinity College, Cambridge

Hardy was born in Cranleigh, in Surrey, and had an enlightened upbringing in a typical Victorian household. He attended Winchester College before proceeding to Trinity College, Cambridge in 1896.

Littlewood was born in Rochester, in Kent. After spending eight years in South Africa, he returned to England and won a scholarship to Trinity College in 1903.

HARDY AND LITTLEWOOD

Hardy's first research paper was on integration, in 1900; he later wrote another sixty-eight papers on the same subject. His textbook, *A Course in Pure Mathematics,* was published in 1908. A model of clarity, it presented elementary analysis to students in a rigorous yet accessible way, and came to have a major impact on English analysis. Also in that year Hardy solved a problem in genetics, using only simple algebra, and sent it to *Science*: 'Hardy's law' has since proved to be of importance in the study of blood groups.

Littlewood's first research paper was on integral functions, in 1906. After spending three years at Manchester University, he returned to Trinity as a College lecturer.

It was in 1912 that they began their remarkable collaboration. Both were geniuses, but Littlewood was probably the more original and imaginative while Hardy was the consummate craftsman, a master of stylish writing. As one of their admirers, the Danish mathematician Harald Bohr, observed:

> *Nowadays, there are only three really great English mathematicians, Hardy, Littlewood, and Hardy–Littlewood.*

THE RAMANUJAN YEARS

In 1913, Bertrand Russell wrote to a friend:

> *In Hall I found Hardy and Littlewood in a state of wild excitement, because they believe they*

have found a second Newton, a Hindu clerk in Madras on £20 a year.

This 'second Newton' was Srinivasa Ramanujan, who had written to Hardy submitting his mathematical discoveries on prime numbers, series and integrals. Although some were incorrect, others showed amazing insight, and Hardy and Littlewood surmised that they must be correct since no-one would have the imagination to make them up. Ramanujan was clearly a genius of the first order, but untutored in formal mathematics.

Srinivasa Ramanujan

Hardy and Littlewood invited him to Cambridge, where they collaborated on several groundbreaking papers. But in 1917, as a result of the climate and a poor diet, Ramanujan contracted tuberculosis. A well-known story tells of Hardy visiting him in hospital: unable to think what to say, Hardy recalled that his taxicab number was *1729 — rather a dull number.* Ramanujan immediately replied:

No, Hardy! It is the smallest number that can be written as the sum of two cubes in two different ways.

Ramanujan returned to India in 1919, and died the following year. Hardy was devastated: he considered Ramanujan to have had the intellectual ability of an Euler or a Gauss.

AFTER RAMANUJAN

By 1919 Hardy felt the need for a break from Cambridge and was appointed Savilian Professor of Geometry in Oxford, where he spent eleven years. There he reformed the mathematics curriculum and built up an impressive school of analysis. His research output blossomed: during his Oxford years he wrote a hundred papers, over half of them with Littlewood who was still in Cambridge. Of this period, Hardy claimed that:

I was at my best at a little past forty, when I was a professor at Oxford.

In 1931 the Sadleirian Chair in Cambridge, Cayley's former chair, fell vacant. Hardy was duly appointed, and spent the rest of his life back at Trinity.

Both Hardy and Littlewood wrote well-known books explaining the nature of mathematics to a general readership. Hardy's *A Mathematician's Apology* (1940) is a personal account by a mathematician looking back as his powers are waning, while Littlewood's *A Mathematician's Miscellany* (1953) is a more joyful work, full of mathematical gems, that allows his readers to experience academic life at Trinity through his perceptive eyes.

Hardy died on the very day that he was due to be presented with the Copley Medal by the Royal Society. His epitaph could well be this sentence from *A Mathematican's Apology:*

I still say to myself when I am depressed, and find myself forced to listen to pompous and tiresome people, "Well, I have done one thing you could never have done, and that is to have collaborated with both Littlewood and Ramanujan on something like equal terms."

Littlewood outlived him by thirty years.

NOETHER

Emmy Noether (1882–1935) was one of the most distinguished mathematicians of the 20th century, contributing to invariant theory, the theory of relativity and, especially, algebra. But being a woman and Jewish, she found herself subjected to great prejudice at several stages of her career.

Emmy Noether as a young woman

Emmy Noether was born in Erlangen in Bavaria, where her father, the algebraist Max Noether, was professor of mathematics. At school she excelled in languages and trained to be a language teacher. But in 1900 she decided to change direction and studied mathematics at the University of Erlangen, where women were allowed to attend classes unofficially with the lecturers' permission. She passed her final university examinations in 1903.

During the following winter she attended the University of Göttingen, where she went to lectures by Hilbert, Klein and Minkowski, but returned to Erlangen where women were now accepted as students. She became officially registered there, and three years later was awarded a doctorate for a thesis in invariant theory.

At this stage she wished to return to Göttingen, but the regulations did not allow women to hold academic positions there. So she remained in Erlangen, helping her ailing father with his teaching commitments, while continuing with her academic research and publishing several papers. Her academic reputation began to spread and she was invited to give a number of prestigious lectures.

MATHEMATICAL PHYSICS IN GÖTTINGEN

In 1915, the year of Albert Einstein's general theory of relativity, Hilbert and Klein invited Emmy Noether back to Göttingen. Hilbert was also working on general relativity, and Noether was welcomed because of her deep knowledge of invariant theory.

Before long, she proved *Noether's theorem,* a cornerstone of general relativity and particle physics, which relates any conservation law in physics to invariance or symmetry properties. On learning of her result, Einstein wrote to Hilbert:

I'm impressed that such things can be

understood in such a general way. The old guard at Göttingen should take some lessons from Miss Noether! She seems to know her stuff.

Meanwhile, Hilbert and Klein were fighting the authorities to permit her to lecture in the University. Although Hilbert enabled her to do so by advertising her lectures under his own name, others, such as the philosophical faculty, were bitterly opposed, exclaiming:

What will our soldiers think when they return to the university and find that they are required to learn at the feet of a woman?

Hilbert angrily replied that the candidate's sex was of no importance, concluding with his memorable riposte:

We are a university, not a bath-house.

The battle was finally won in 1919.

ALGEBRA

Noether continued to write papers on invariant theory and relativity until 1920, when she had a change of direction and became interested in algebra — and in particular, the study of commutative rings. It is for her work in this area that she is best remembered.

Earlier we met the idea of a *group*, an algebraic object that consists of a set of elements and a single way of combining them in pairs so as to satisfy certain specified rules. Another algebraic object of interest is a *ring*, which consists of a set of elements and *two* ways of combining them in pairs so as to satisfy certain specified rules. Examples of rings include:

- adding and multiplying integers to give other integers
- adding and multiplying complex numbers to give other complex numbers
- adding and multiplying polynomials to give other polynomials
- adding and multiplying matrices (rectangular arrays of numbers) to give other matrices.

If, moreover, multiplication is commutative — that is, $a \times b = b \times a$, for all elements a and b in the set — then we have a *commutative ring*. The first three of the above rings are commutative, but not the last one.

In 1921 Emmy Noether wrote a classic paper, *Idealtheorie in Ringbereichen* (Theory of ideals in ring domains), in which she investigated the internal structure of commutative rings in terms of certain subsets called *ideals*. In particular, she studied rings in which a particular property of these ideals holds, and such rings are now known as *Noetherian rings*. Her researches in algebra continued throughout the 1920s and were rewarded with invitations to speak at the International Congresses of Mathematicians in Bologna in 1928 and Zürich in 1932.

LEAVING GERMANY

In 1933, with the rise of Adolf Hitler, the Nazis withdrew the right of Jews to teach at the University, and she was forced to leave Germany and seek employment elsewhere.

Eventually she was given a position at Bryn Mawr, a women's college in the USA, near Philadelphia, and was also invited to lecture at the Institute for Advanced Study in Princeton. She was blissfully happy at Bryn Mawr, with congenial colleagues, but less than two years after arriving there she developed a large ovarian cyst and died.

Bryn Mawr College

VON NEUMANN

The enormous range of interests of John von Neumann (1903–1957) is remarkable. He worked on the foundations of set theory and quantum mechanics, developed the algebra of operators on a Hilbert space, and founded the subject of game theory. His work on mathematical physics, particularly in turbulence, detonation waves and shocks in fluids, was very influential. He advanced the theory of cellular automata, and with his introduction of the stored program concept is often called 'the Father of Modern Computing'.

Von Neumann was born in Budapest where, in 1926, he received his doctorate with a thesis on set theory. By his mid-20s, he had an international reputation in the academic community. He lectured at Berlin and Hamburg until 1930, and for part of that time also studied with Hilbert at Göttingen. He then taught for three years at Princeton University until he was appointed as one of the founding professors at the newly created Princeton Institute for Advanced Study, a position he held for the rest of his life.

Oskar Morgenstern

During and after the Second World War, von Neumann was an adviser to the American military on weapons development, in particular on atomic weapons and logistics. From 1943 to 1955 he was a consultant to the Los Alamos Scientific Laboratory. It has been suggested that the cancer that caused his death may have arisen from his being a witness at atomic bomb testings.

GAME THEORY

Von Neumann's work on games is characteristic of a lifelong approach of using mathematics in practical situations. The consequences of his work go far beyond its applications to games of chance, such as poker, and have been important in psychology, sociology, politics and military strategy. His 1944 book with his Princeton colleague, Oskar Morgenstern, revolutionized the field of economics.

A *zero-sum game* between two players is one in which the gain to the winning player exactly equals the loss to the loser, so the total payoff to both players sums to zero. In 1928 von Neumann published his *minimax theorem* which proves, for a zero-sum two-person game, that both participants have strategies (or methods of playing) that minimize their maximum loss. As he remarked:

> *As far as I can see, there could be no theory of games ... without that theorem ... I thought there was nothing worth publishing until the 'Minimax Theorem' was proved.*

His theory was later expanded to cover more general situations.

COMPUTING

After the war von Neumann led a team at Princeton in the development of a computer. They decided that it should have four main components:

- An arithmetic/logic unit, now called the central processing unit (CPU), which is where the basic elementary operations are performed: this was analogous to Babbage's mill.

John von Neumann (right) and Robert Oppenheimer with the EDVAC computer

- A memory (analogous to Babbage's store) that stored the numbers on which the calculations were to be performed and also the instructions for performing the calculations. Since these instructions could be coded as numbers, the machine needed to be able to distinguish between numbers and coded instructions.
- A control unit that decoded and executed the instructions fetched from the memory.
- Input/output devices to allow data and instructions to be entered into the computer and the results of the calculations displayed. Von Neumann was particularly interested in output devices that could display results graphically.

Because it used electronic technology, numbers in the machine were represented in binary form, so that any device holding a digit needed to have only two states.

The machine was completed in 1952. It had 3600 vacuum tubes and was the first stored program computer, unlike previous computers that were programmed by altering their circuitry. Although it had its drawbacks, von Neumann's design model was very influential in the subsequent development of computers.

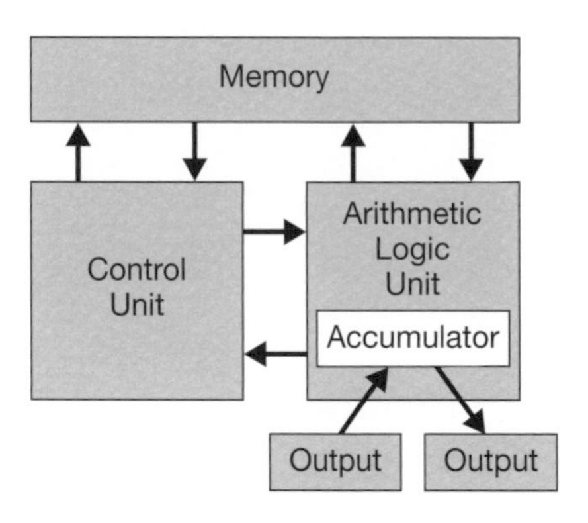

Von Neumann design architecture

TURING

Alan Turing (1912–1954) was a mathematician, logician, philosopher, codebreaker, and a founder of computer science. He is remembered for the *Turing machine* which formalized the ideas of algorithm and computation, and the *Turing test* in artificial or machine intelligence. His codebreaking activities during the Second World War attacked the operation of the German cipher machines, with a subtle analysis of their logical structure.

Turing was born in London and entered King's College, Cambridge, in 1931, being elected to a fellowship there after graduating. In 1936 he went to Princeton University to pursue doctoral studies, but returned to Cambridge in 1938.

When war was declared, Turing moved to the Government Code and Cypher School at Bletchley Park. After it was over, he went to the National Physical Laboratory in London to work on the design of an electronic computer, the Automatic Computing Engine (ACE). His final university position was as Deputy Director of the Manchester Computing Laboratory.

Turing continued to be consulted by GCHQ, the successor to Bletchley Park, but lost his security clearance when he was brought to trial for homosexual activities in 1952. He came under intense scrutiny by the intelligence services, who regarded him as a security risk. He died of cyanide poisoning, a half-eaten apple beside his bed, and the inquest brought in a verdict of suicide.

TURING MACHINE

Turing was intrigued by Hilbert's decision problem, a problem in mathematical logic:

Given a mathematical proposition, can one find an algorithm to decide whether the proposition is true or false?

To tackle this, Turing needed a workable definition of an algorithm, and he identified this with the output of an abstract machine, later called a *Turing machine,* which consisted of an infinite tape and a component that could be in any one of a finite number of states. These states could change, depending on the current symbol that was read from the tape.

He then went a step further and envisaged a *universal Turing machine* that could emulate all other Turing machines — an analogy is the modern computer which can do different tasks if its programming is altered appropriately. In 1936 Turing answered the question in the negative: there are mathematical propositions that are *undecidable* — no algorithm can decide whether they are true or false. The idea of a Turing machine became the foundation of the theory of computation.

BLETCHLEY PARK

Bletchley Park was the home of the British codebreaking efforts in the Second World War. There, Turing and his colleagues attacked the codes generated by the German electro-mechanical rotor-coding machines *Enigma* and *Lorenz.*

In the example below, when a key is pressed, a current flows through the three rotors, is reflected, and then flows back through the rotors to light up a lamp to give the code for the pressed key. Then, crucially, one of the rotors rotates, creating a new pathway for the signals. When the first rotor has made a complete turn, as a result of key presses, the middle one starts to move, and when it makes a complete turn the last one starts to move.

A logical feature of this machine is that, in any given state of the rotors, the coding is symmetrical: for example, if letter Q is coded as

GROWTH AND FORM IN BIOLOGY

Turing had a lifelong interest in the development of pattern and form in living organisms. Towards the end of his life he applied various mathematical techniques to the subject. He particularly wanted to explain the appearance of Fibonacci numbers in plants — for example, in the spiral patterns in sunflower heads.

He first looked at how biological systems that are symmetrical at the start can lose that symmetry, and wondered how this could be caused by the dynamics of the way that the chemicals diffuse and react. Using a computer, he carried out pioneering work in modelling these chemical reactions, and published his results in his 1952 paper, *The chemical basis of morphogenesis.*

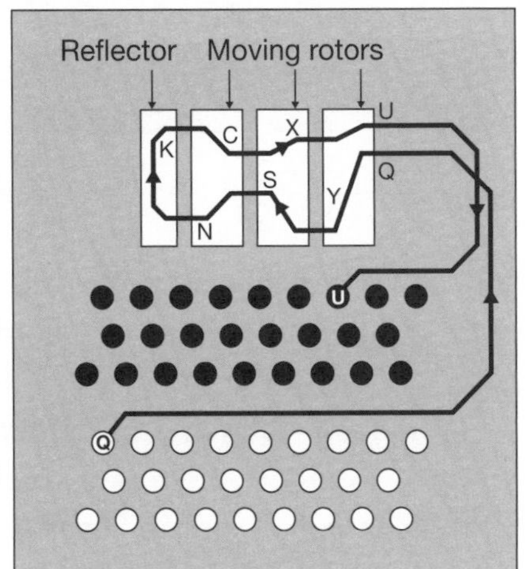

letter U, then U would be coded as Q; in particular, it can never code a letter into itself. Turing used such an analysis of the logical structure of the machine to allow the successful decoding of messages.

The official historian of British Intelligence in World War II has said that the intelligence provided by Bletchley Park shortened the war 'by not less than two years and probably by four years'.

THE TURING TEST

In his 1950 paper, *Computing Machinery and Intelligence,* in the journal *Mind,* Turing starts

> *I propose to consider the question, 'Can machines think?'*

He refined this question by having a machine, a woman and an interrogator, with the interrogator in a different room from the other two. The object of the game, called the *Turing test,* is for the interrogator to determine which of the others is the machine and which is the woman. They can communicate with the interrogator, but only in a manner that gives no clue to their identities.

In his previous work with Turing machines, he concentrated on what machines cannot do. Now his focus was on what they can do, and in particular whether the behaviour of the brain can be replicated by a computer. The Turing test remains important in philosophy and artificial intelligence.

Alan Turing

BOURBAKI

Charles Bourbaki was a 19th-century French general who distinguished himself in the Crimean and Franco-Prussian Wars but is not known to have had an interest in mathematics. His namesake, Nicolas Bourbaki (b. 1934), was a French mathematician who never existed, but whose writings profoundly influenced much of pure mathematics in the 20th century!

Two Bourbakists: Henri Cartan and Jean-Pierre Serre

Bourbaki was the pseudonym of a group of (mainly French) mathematicians who over a period of thirty years produced an influential series of books that were designed to present all of pure mathematics in a completely structured and axiomatic way.

THE BIRTH OF 'BOURBAKI'

In late 1934, André Weil and Henry Cartan were complaining about a calculus textbook they were required to teach from. As regular participants of a mathematics seminar series in Paris, they used to meet their colleagues over lunch in the Café Capoulade in Paris's Latin Quarter to discuss matters of mathematical interest. It was at one such lunchtime gathering that several of these friends agreed to form a group to write a superior calculus text. Their original plan was for a single 1000-page book that started from first principles and developed the subject systematically, with each result depending on earlier ones and nothing taken for granted.

In addition to Weil and Cartan, the Bourbaki group's founder members included Claude Chevalley, Jean Delsarte, Jean Dieudonné and René de Possel, with others added shortly after.

> **BOURBAKI'S FIRST SIX BOOKS**
> *Book I:* Set Theory
> *Book II*: Algebra
> *Book III*: Topology
> *Book IV*: Functions of One Real Variable
> *Book V*: Topological Vector Spaces
> *Book VI*: Integration

THE FIRST BOOKS

While discussing how to structure their calculus book coherently, they realised the need to go back to basics and tidy up the foundations of their subject. They soon agreed that a whole series of books was needed, and the project became ever more ambitious as they sought to lay down ground rules for the whole of pure mathematics.

For the title of the series, they chose the all-encompassing *Éléments de Mathématique.* The first book to appear was on set theory, published in 1939. The Second World War then intervened. Individual chapters of the various books continued to be written, but it was not until 1958 that the next five books were completed.

Writing the books was no easy task. The Bourbaki group members met for one or two weeks at a time, and arguments were frequent as they battled over the best way to structure the

WHY 'BOURBAKI'?

Around 1918, André Weil and the other first-year students at the École Normale Supérieure in Paris were invited to a mathematics lecture at which the 'distinguished lecturer' (actually, a senior student) appeared with a false beard in the guise of a celebrated mathematician and presented a number of 'theorems', each more ridiculous than the previous one and each (for some reason) named after a 19th-century French general. The lecturer named his final result *Bourbaki's theorem*, after General Bourbaki. Recalling this spoof lecture several years later, Weil mischievously proposed that the group should adopt the pseudonym 'Bourbaki' as the author of all their books.

General Charles Bourbaki

foundations of pure mathematics so as to influence its future development. Presenting the material abstractly was all-important, leading to a highly abstract treatment in which rigour and structure were paramount — even our basic number system did not feature until all the foundations had been firmly laid.

(Back row) Cartan, de Possel, Dieudonné, Weil, lab technician; (seated left to right) Mirlès, Chevalley, Mandelbrojt

THE 'DECLINE' OF BOURBAKI

The first six books proved to be very influential, but they also had many detractors. They were not textbooks in the traditional sense: their contents did not progress from simple ideas to more complicated ones, and the highly structured material was difficult to read. Moreover, applications were ignored and problem solving was de-emphasized.

The heyday of the Bourbaki books was in the 1950s and 1960s, as new members joined the group to continue the project: these included Pierre Samuel, Jean-Pierre Serre and Laurent Schwarz, and later, Armand Borel, Serge Lang, Alexandre Grothendieck and John Tate, all with distinguished research careers in their own right.

Eventually the project began to founder as mathematics expanded and headed in diverse directions and the team was unable to keep up. But despite all its problems, the Bourbaki group refuses to lie down and new editions of its books continue to appear.

ROBINSON AND MATIYASEVICH

One of Hilbert's twenty-three problems, Problem 10, asked whether there is a systematic procedure for deciding whether a certain type of equation has any solutions that are integers. The key to solving this problem was a paper by Julia Robinson (1919–1985) from the USA, and the answer was given by a young Russian mathematician, Yuri Matiyasevich (b.1947). The fruitful collaboration between them was conducted across two continents.

Julia Robinson (née Bowman) had a troubled childhood. Her mother died when she was 2 and she and her sister were sent to live in a small community in the Arizona desert. When her father remarried, they moved to San Diego, but she became ill with scarlet fever and rheumatic fever and missed two years of schooling — and from then on, her health was never good. After catching up with her studies, she became interested in mathematics and physics, which she studied at San Diego State College, later transferring to the University of California in Berkeley.

At Berkeley she attended lectures on the history of mathematics, where she was much inspired by E. T. Bell's *Men of Mathematics,* and on number theory by Raphael Robinson, who later became her husband. After her graduation she worked for a doctorate on mathematical logic with the logician Alfred Tarski. Although she wrote an influential paper on two-person zero-sum games (while working for a year for the RAND Corporation) and a paper on statistics, almost all of her subsequent publications were connected with Hilbert's Problem 10.

Julia Robinson

HILBERT'S PROBLEM 10

We have seen that a Diophantine equation is an equation for which we seek solutions that are integers; examples of these include the Pythagorean theorem $x^2 + y^2 = z^2$ (one of whose integer solutions is $x = 3$, $y = 4$, $z = 5$), Pell's equation $3x^2 + 1 = y^2$ (one of whose integer solutions is $x = 1$, $y = 2$), and Fermat's theorem $x^4 + y^4 = z^4$ (which has no positive integer solutions).

Hilbert's Problem 10 was concerned with determining whether Diophantine equations have integer solutions, such as the first two above:

> *Given a Diophantine equation with integer coefficients, does there exist a finite step-by-step procedure for deciding whether the equation has integer solutions?*

This problem is entirely existential — it does not ask how to find such solutions if they exist.

If the answer to Hilbert's problem is *yes,* then this can be decided by producing a specific procedure (called an *algorithm*). But if the answer is *no,* then the problem becomes much harder, as one has to prove that no such algorithm can possibly exist.

YURI MATIYASEVICH

Robinson's main contribution to the problem was what became known as the *'Robinson hypothesis'*: to show that no such algorithm exists, find (in her words)

> *some diophantine relation that grows faster than a polynomial, but not too terribly fast*

— for example, the powers of 2.

SEQUENCE-GENERATING POLYNOMIALS

One of the consequences of the results of Robinson and Matiyasevich is that one can prove the existence of polynomials (in several variables) that can take specified values when non-negative integers are substituted for the variables. For example, consider the polynomial

$$2xy^4 + x^2y^3 - 2x^3y^2 - y^5 - x^4y + 2y.$$

As x and y range over all the non-negative integers, this expression gives both positive and negative values. But *all the positive ones are Fibonacci numbers and every Fibonacci number can be obtained in this way;* for example, choosing $x = 5$ and $y = 8$ gives the Fibonacci number 8.

Similarly, consider the following polynomial in twenty-six variables, $a, b, \ldots, z$:

$$\begin{aligned}
&(k+2)\{1 - [wz + h + j - g]^2 + [(gk + 2g + k + 1)(h + j) + h - z]^2 \\
&- [16(k+1)^3(k+2)(n+1)^2 + 1 - f^2]^2 - [2n + p + q + z - e]^2 \\
&- [e^3(e+2)(a+1)^2 + 1 - o^2]^2 - [(a^2 - 1)y^2 + 1 - x^2]^2 - [16r^2y^4(a^2 - 1) + 1 - u^2]^2 \\
&- [((a + u^2(u^2 - a))^2 - 1)(n + 4dy)^2 + 1 - (x + cu)^2]^2 - [(a^2 - 1)l^2 + 1 - m^2]^2 \\
&- [ai + k + 1 - l - i]^2 - [n + l + v - y]^2 \\
&- [p + l(a - n - 1) + b(2an + 2a - n^2 - 2n - 2) - m]^2 \\
&- [q + y(a - p - 1) + s(2ap + 2a - p^2 - 2p - 2) - x]^2 \\
&- [z + pl(a - p) + t(2ap - p^2 - 1) - pm]^2\}.
\end{aligned}$$

As $a, b, c, \ldots$, range over all the non-negative integers, this expression gives both positive and negative values. But *all the positive ones are prime numbers, and every prime number can be obtained in this way.*

Another mathematician who became interested in Hilbert's Problem 10 was the young Russian, Yuri Matiyasevich, from St Petersburg. While still an undergraduate, he had been given the problem as a project to work on. In 1970, at the age of 22, he used Robinson's hypothesis to solve Hilbert's problem in the negative — there exists no such algorithm — but instead of the powers of 2, he used the Fibonacci numbers.

Yuri Matiyasevich

Robinson and Matiyasevich eventually developed their ideas in several joint papers. But with no photocopiers, and with their lengthy long-hand letters taking three weeks to arrive (and occasionally being lost in the post), communications were difficult. Visa problems frequently prevented Matiyasevich from travelling, but they eventually met at a conference in Bucharest, Romania, in 1971, and again in Calgary, Canada, in 1982.

After her major role in solving Hilbert's problem, Julia Robinson was much sought after, receiving many prestigious invitations. She gave the American Mathematical Society's Colloquium Lectures, and was the first woman to be appointed to the U.S. National Academy of Sciences and to become President of the American Mathematical Society.

APPEL AND HAKEN

How many colours are needed to colour a map on the plane so that neighbouring countries are coloured differently? This problem was first raised in 1852, but was not solved until 1976. Its method of solution by Kenneth Appel (b. 1932) and Wolfgang Haken (b. 1928) was controversial, as its heavy reliance on a computer raised philosophical questions about the nature of mathematical proof.

On 23 October 1852 Augustus De Morgan, professor of mathematics at University College, London, wrote to William Rowan Hamilton:

A student of mine asked me today to give him a reason for a fact which I did not know was a fact — and do not yet. He says that if a figure be any how divided and the compartments differently coloured so that figures with any portion of common boundary line *are differently coloured — four colours may be wanted, but not more ... Query cannot a necessity for five or more be invented ... My pupil says he guessed it in colouring a map of England ... .*

But De Morgan was unable to prove what became known as the *Four-colour theorem:*

Every map can be coloured with at most four colours.

After De Morgan's death, the problem was largely forgotten until Arthur Cayley raised it at a meeting of the London Mathematical Society in June 1878. In the following year, his former Cambridge student Alfred Kempe produced a 'proof' that was widely accepted until Percy Heawood of Durham found a fatal error in it in 1890, but salvaged enough from Kempe's proof to deduce that *every map can be coloured with at most five colours.*

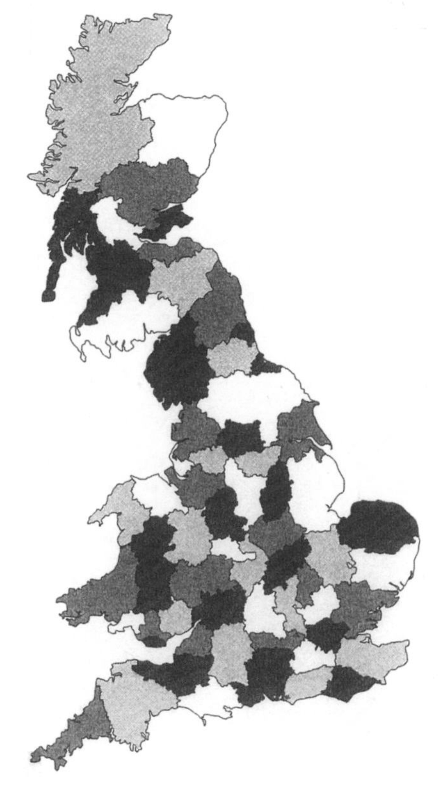

A four-coloured map of Britain

TWO IMPORTANT IDEAS

Although Kempe's proof was faulty, it contained two useful ideas that would feature prominently in later attempts and in the eventual solution by Appel and Haken.

- **Unavoidable sets:** Kempe proved that *every map must contain at least one country bounded by two, three, four or five boundary lines* — that is, a digon, triangle, quadrilateral or pentagon. Such a set of four configurations is called *unavoidable* since no map can avoid at least one of them.
- **Reducible configurations:** He also proved that every map that contains a digon, triangle or quadrilateral can be four-coloured, so none of these can appear in a counter-example: we call such configurations *reducible.* Kempe believed that he had also proved pentagons to be reducible, but here his argument was faulty — if it had been correct, the Four-colour theorem would have been proved.

In 1904 it was proved that if a map contains no digon, triangle or quadrilateral, then not only must it contain a pentagon, but it must contain either two adjacent pentagons or a pentagon adjacent to a hexagon. Thus, the set consisting of a digon, triangle, quadrilateral and these two configurations is also an unavoidable set.

In another direction, the distinguished American mathematician George Birkhoff, who had became fascinated by the four-colour problem, proved that the 'Birkhoff diamond' of four adjacent pentagons is also reducible (see below).

From this point, the search was on for unavoidable sets and reducible configurations. One map-colourer who found several of the former was Henri Lebesgue, better known to mathematicians for his 'Lebesgue integral', while over the years many thousands of the latter were discovered.

The aim of such a search, as pointed out by the German mathematician Heinrich Heesch, who had solved one of Hilbert's problems and who himself came near to proving the Four-colour theorem, was to produce *an unavoidable set of reducible configurations.* Finding such a set proves the theorem: since the set of configurations is unavoidable, every map must contain at least one of them, but since each configuration in the set is reducible, the colouring of the map can be completed in every case.

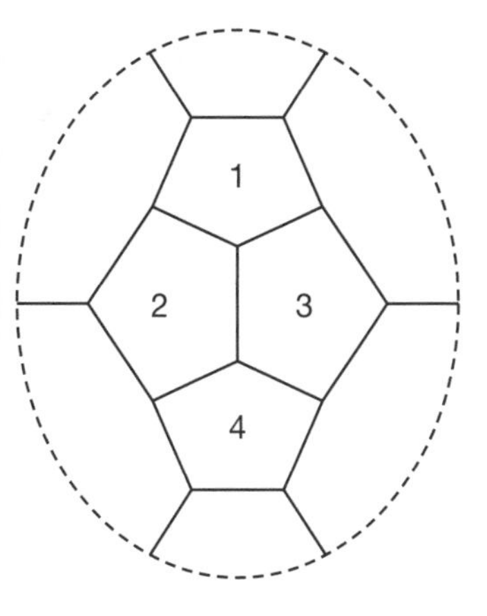

George David Birkhoff and his 'diamond'

Kenneth Appel and Wolfgang Haken

APPEL AND HAKEN'S SOLUTION

The Four-colour theorem was eventually proved by Kenneth Appel and Wolfgang Haken of the University of Illinois. They announced their proof on 24 July 1976.

Haken had been introduced to the problem by Heesch, who believed there to be a finite (though very large) unavoidable set of reducible configurations. But whereas most investigators had generated large numbers of reducible configurations and then tried to package them into an unavoidable set, Appel and Haken's approach was to seek unavoidable sets of 'likely-to-be-reducible' configurations and then use the computer to test these for reducibility, modifying the set as necessary. This massive task, involving some twelve hundred hours of computer time and a three-year man-computer dialogue, eventually led to an unavoidable set of 1936 reducible configurations, later reduced to 1482.

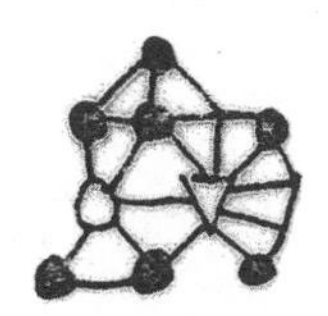
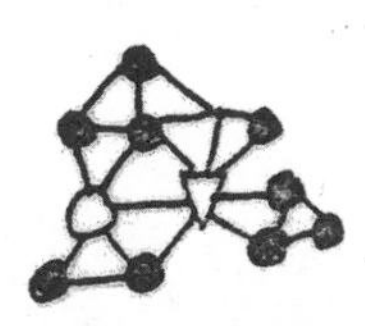
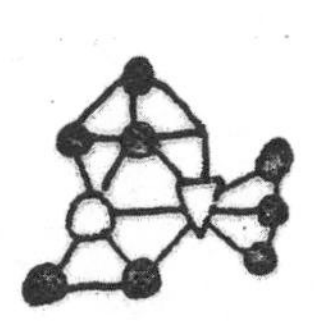

Three of Appel and Haken's configurations

MANDELBROT

An exciting area of mathematics that emerged in the 20th century, although it can trace its origins back to Bolzano and Poincaré, is that of fractal geometry. This topic was espoused and greatly developed by Benoit Mandelbrot (1924–2010), and is closely associated with the fashionable area of chaos theory.

Mandelbrot was born in Poland and spent most of his working life at the computer firm IBM in the USA. He joined Yale University at the age of 75 and retired in 2005.

How long is the coastline of Britain? If you try to measure it with a ruler, or if you look at the country from far above the earth, you can estimate the length of the coastline. But as you measure more accurately, or get closer to the earth, you become aware of more and more inlets and bays, and the length increases accordingly. The closer you get, the longer the coastline seems to become. In fact, the coastline of Britain has infinite length, even though it encloses a finite area.

Benoit Mandelbrot

VON KOCH'S SNOWFLAKE CURVE

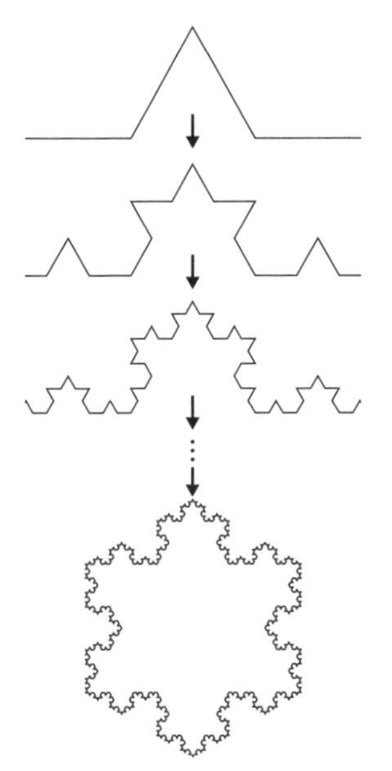

A similar situation occurs with the *snowflake curve,* described in 1906 by the Swedish mathematician Helge von Koch. To construct it, take an equilateral triangle, and then replace the middle third of each side (which we can think of as the base of a smaller equilateral triangle) by the other two sides of the triangle, giving a 'peak' on each side of the original triangle. Now repeat this process with each of the lines in the resulting picture. Carrying on this process for ever gives the snowflake curve.

Like the coastline of Britain it has infinite length, yet encloses a finite area. It is also *self-similar* — parts of it have the same shape (though smaller) when you look at it in closer detail. This self-similarity is a standard feature of fractal patterns, a topic of great interest in the 20th century.

THE MANDELBROT SET

A new way of obtaining fractal patterns was described by Benoit Mandelbrot. Consider the transformation $z \rightarrow z^2 + c$, where c is a fixed complex number. For each initial value, square it and add c to give a new number, and continually repeat the process. For example, when $c = 0$ and $z \rightarrow z^2$:

- the initial value 2 gives 4, 16, 256, ... , going off to infinity;
- the initial value $\frac{1}{2}$ gives $\frac{1}{4}$, $\frac{1}{16}$, $\frac{1}{256}$, ... , which tends to 0.

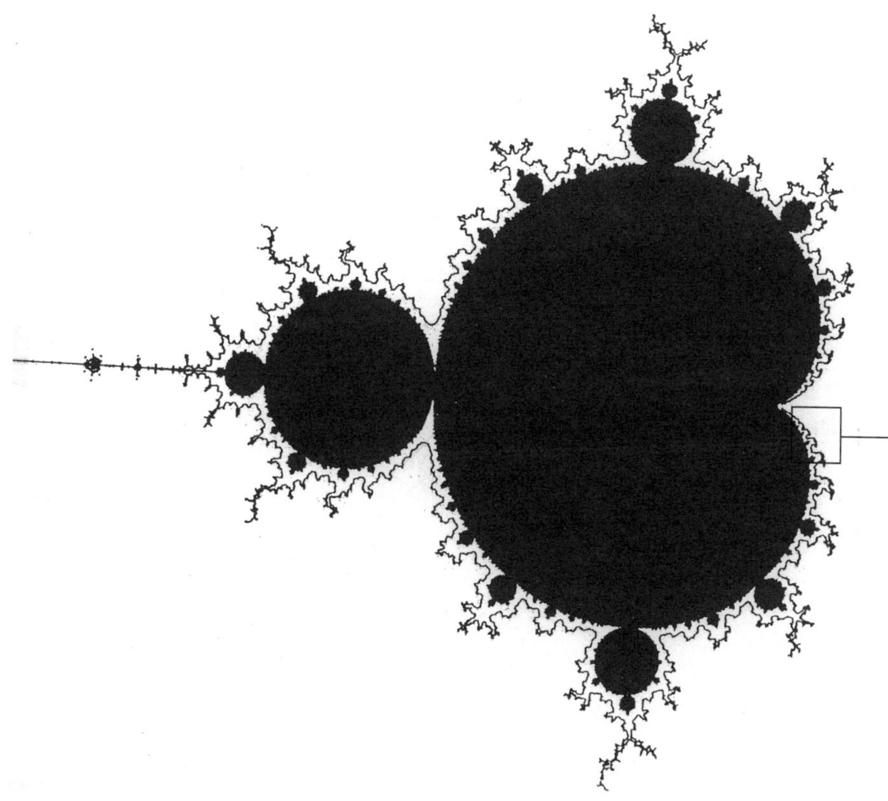

The Mandelbrot set and a detail from its boundary

Here, all points inside the circle with centre 0 and radius 1 stay inside the circle, all points on the circle stay on the circle, and all points outside the circle goes to infinity.

1

0

$c = 0$

We call this boundary circle the *Julia set for* $c = 0$ (after the French mathematician Gaston Julia), and its inside is the *keep set* (because we keep its points in sight).

Different values of c give a wide range of different boundary curves (Julia sets): for example, $c = 0.25$ gives us a 'cauliflower' shape, while $c = -0.123 + 0.745i$ gives us a shape somewhat like a rabbit. The Julia sets for some values of c are in one piece, while others are in several pieces.

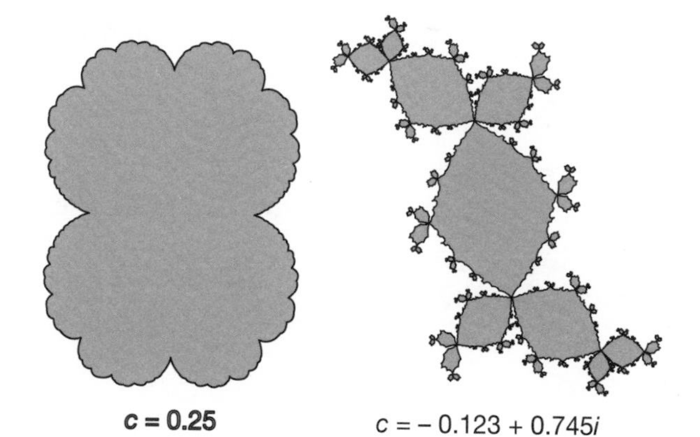

Two Julia sets

Mandelbrot drew a picture of all the complex numbers c for which the Julia set is in one piece, obtaining a fascinating picture now called the *Mandelbrot set.* This set arises in the study of chaos theory, where it shows how sensitive the behaviour of the transformation can be on the choice of the number c. It has given rise to a whole range of beautiful designs under the heading of *fractal art.*

WILES

It is not given to everyone to achieve their childhood dream, but this happened for Sir Andrew Wiles (b.1953), who encountered Fermat's last theorem as a schoolboy, worked on it single-mindedly for many years, and eventually proved it after a lengthy and difficult struggle and a magical moment.

Wiles recalls when he first came across Fermat's last theorem:

One day I happened to be looking in my public library and I found a book on maths — it told a bit about the history of this problem and I, a ten year old, could understand it. From that moment I tried to solve it myself — it was such a challenge, such a beautiful problem — this problem was Fermat's last theorem.

Wiles studied at Oxford University and then took his doctorate in Cambridge. He spent some time at Harvard University and in Germany before taking up an appointment at the Institute for Advanced Study in Princeton, where he spent almost twenty years. He has now returned to Oxford University.

FERMAT'S LAST THEOREM

Earlier, we saw that Fermat proved that the equation $x^4 + y^4 = z^4$ has no positive integer solutions x, y and z, and that Euler proved a similar result for the equation $x^3 + y^3 = z^3$. But 'Fermat's last theorem', which asserts that

For any integer n (> 2), there are no positive integers x, y and z for which $x^n + y^n = z^n$,

remained unproved for all larger values of n.

In order to prove Fermat's last theorem in general, it is enough to prove it when n is a prime number — for example, we can reduce the case $n = 20$ to the case $n = 5$ by writing

$$x^{20} + y^{20} = z^{20} \quad \text{as} \quad X^5 + Y^5 = Z^5,$$

where $X = x^4$, $Y = y^4$ and $Z = z^4$.

In the 19th century, proofs were found for $n = 5$ and $n = 7$, and the German number-theorist Ernst Kummer proved it also for a large class of primes called 'regular primes'. Much later, building on his work, and making extensive use of modern computers, the list was extended to all primes below 4,000,000.

THE BREAKTHROUGH

Two ideas that were central to Wiles's eventual proof are those of an 'elliptic curve' and a 'modular form'. An *elliptic curve* is essentially a curve whose equation is of the form

$$y^2 = x^3 + rx^2 + sx + t,$$

for some integers r, s and t. A *modular form* may be thought of, in general terms, as a way of generalizing the Möbius transformation

$$f(z) = (az + b) / (cz + d).$$

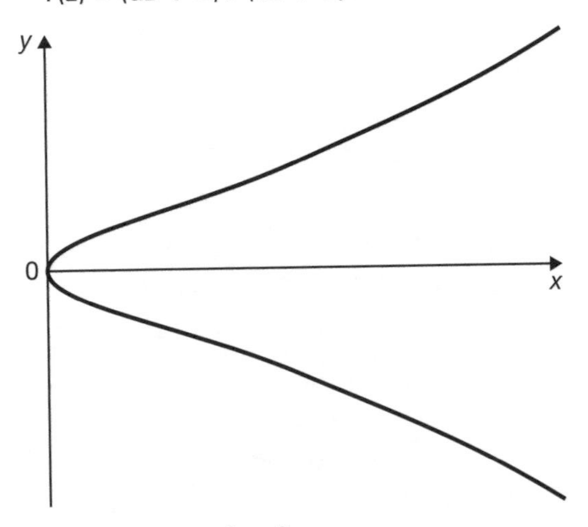

The elliptic curve $y^2 = x^3 + x$

Yutaka Taniyama and Goro Shimura conjectured that *Every elliptic curve is associated with a modular form*, and it was quickly realized that a proof of this conjecture (or at least a special case of it) would imply the truth of Fermat's last theorem.

Around 1984 major progress was made when

Andrew Wiles lectures on elliptic curves

Gerhard Frey of Saarbrücken spotted that if Fermat's theorem were false, so that the equation $a^p + b^p = c^p$ held for some positive integers a, b and c and prime number p, then the elliptic curve

$$y^2 = x^3 + (b^p - a^p)\,x^2 - a^p b^p\,x$$

would have such bizarre properties that it could not be modular, thereby contradicting the Taniyama–Shimura conjecture.

THE PROOF

At this stage, Andrew Wiles entered the fray. Excited by Frey's observations, he set himself the task of proving the special case of the Taniyama–Shimura conjecture that implied the truth of Fermat's last theorem. For seven years he hid himself away from other distractions while he concentrated on chipping away at the problem:

> *You cannot focus yourself for years unless you have this kind of undivided concentration which too many spectators would destroy.*

By 1993 he had convinced himself that he had completed the proof, and he presented it, to world-wide enthusiasm and acclaim, at a major conference at Cambridge University.

But during the detailed checking of the proof by the great and the good, a serious gap was discovered. For over a year, Wiles and his former doctoral student, Richard Taylor, struggled to close the gap. Wiles was about to give up when:

> *Suddenly, unexpectedly, I had this incredible revelation. It was the most important moment of my working life.*
>
> *Nothing I ever do again ... it was so indescribably beautiful, it was so simple and so elegant, and I just stared in disbelief for twenty minutes, then during the day I walked around the department. I'd keep coming back to my desk to see it was still there — it was still there.*

The proof was indeed complete, and Andrew Wiles was able to look back with pride and pleasure on his monumental achievement:

> *There is no other problem that will mean the same to me. I had this very rare privilege of being able to pursue in my adult life what had been my childhood dream. I know it's a rare privilege, but I know if one can do this, it's more rewarding than anything one can imagine.*

PERELMAN

In August 2000, a century after Hilbert had introduced his twenty-three problems in Paris, seven 'millennium problems' were announced, in order to celebrate mathematics in the new millennium. These 'Himalayas of mathematics', each carrying a reward of one million dollars for its solution, were considered by the mathematical community to be the most difficult and important in the subject. One of these is the Riemann hypothesis, which remains unsolved. Another was the *Poincaré conjecture,* posed by Poincaré in 1904, which was recently solved by the Russian mathematician Grigori Perelman (b. 1966).

Topology, sometimes referred to as 'bendy geometry', is the branch of geometry in which two shapes are regarded as the same whenever we can bend, or deform, either into the other. For example, the sphere and the torus are not the same, since we cannot deform either into the other, whereas we can deform a sphere into a cube, or a torus into a teacup. Indeed, a topologist has been described as someone who cannot tell the difference between a bagel and a teacup!

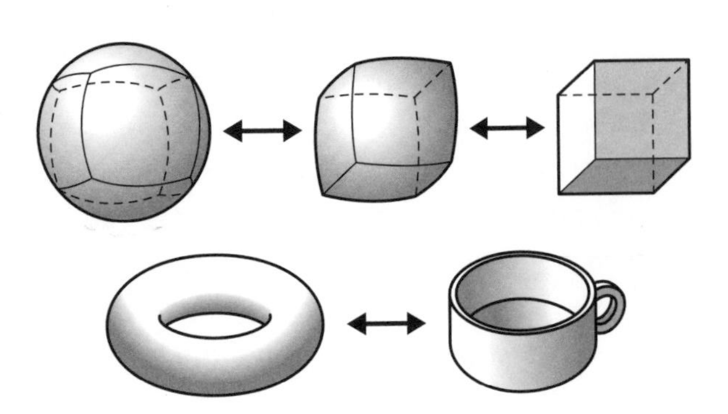

Another way of explaining why the sphere and the torus are not the same is to lay a loop of thread on the surface and try to shrink it to a point. With the sphere you can place the loop anywhere and it will always shrink to a point. On the torus you sometimes can – but not always, because the hole might get in the way.

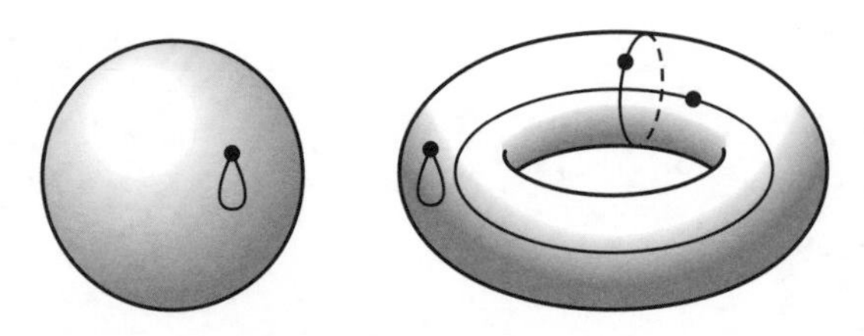

We now restrict our attention to the surfaces of spheres, but vary the dimension:

- We consider a circle to be one-dimensional, even though it lives in two-dimensional space, because it is a 'bent-around line': the equation of such a 'one-dimensional sphere' is $x^2 + y^2 = r^2$, where r is the radius.

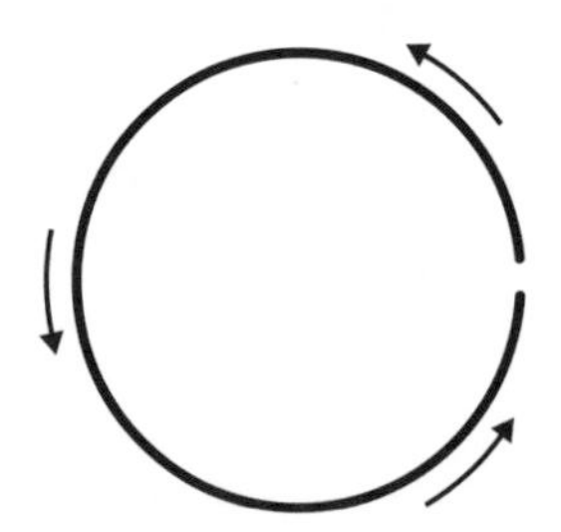

- If we now fill in two circles, bend them, and glue corresponding points together, we get a sphere (see below). We consider this sphere to be two-dimensional, even though it lives in three-dimensional space – think of standing on the surface of the earth and viewing the two-dimensional world around you: the equation of such a 'two-dimensional sphere' is $x^2 + y^2 + z^2 = r^2$.

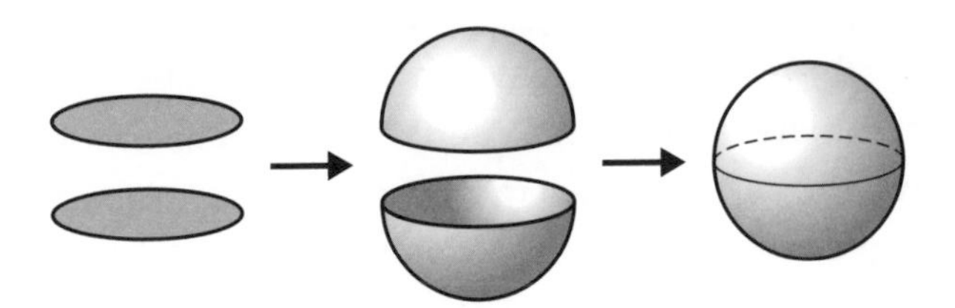

- We can continue in this way. To get a three-dimensional sphere, we take two solid spheres and glue corresponding points together: although this cannot be done, or visualized, in our three-dimensional world, it can still be studied mathematically, giving us an object that lives in four-dimensional space: the equation of such a 'three-dimensional sphere' is $x^2 + y^2 + z^2 + w^2 = r^2$.
- In the same way, we can consider four-dimensional, five-dimensional, and even higher-dimensional, spheres.

THE POINCARÉ CONJECTURE

In two dimensions, it can be proved that

> *The surface of a sphere (or any surface that can be deformed into a sphere) is the only surface with the loop-shrinking property*

– the torus does not have this property, as we saw, and nor does any other type of surface. But what happens in higher dimensions?

> *Is the surface of a higher-dimensional sphere the only surface with the loop-shrinking property?*

Poincaré conjectured that the answer to the last question is *yes*, and this became known as the *Poincaré conjecture.* For the two-dimensional sphere the answer is *yes*, as we have seen, and in the 1960s the American mathematician Stephen Smale proved that the answer is also *yes* for surfaces of dimension 5 or higher. Then, in 1982, another American mathematician, Michael Freedman, proved it for surfaces of dimension 4.

Grigori Perelman

But what about surfaces of dimension 3? This seemed to be the hardest case. No-one could prove it, and in this form it became one of the seven millennium problems.

PERELMAN'S SOLUTION

Perelman was a mathematical prodigy. In his student days he participated in the USSR team in the International Mathematical Olympiad, where he achieved a perfect score.

In 2002 and 2003 he published several papers in which he succeeded in proving the Poincaré conjecture. His solution was very difficult to understand, even for specialists. To analyse the surfaces concerned, he essentially looked at the various ways in which a liquid can flow over them, and studied the dynamics of these so-called 'Ricci flows'; these are somewhat like the flow of heat through a solid object.

His papers caused a sensation throughout the mathematical world. But Perelman shuns publicity, believing passionately that mathematics should be studied for its own sake alone and not for any financial reward. Smale and Freedman both received Fields Medals for their contributions to the solution of the Poincaré conjecture, but when Perelman was offered one at the International Congress of Mathematicians in 2006, he turned it down. Four years later he turned down the million-dollar prize offered to him for solving one of the millennium problems.

FIELDS MEDALLISTS

A special feature of the International Congresses of Mathematics, held every four years, is the award of Fields medals to the most outstanding young mathematicians. For many years these were regarded as the mathematical equivalent of Nobel prizes, but recently a new prize, the Abel Prize, has been instituted and is awarded annually.

The Proceedings of the first International Congress in Zürich, 1897

As part of the 400th anniversary celebrations of Columbus's voyage to America, a 'World Congress' of Mathematicians took place at the World's Columbian Exposition in Chicago in 1893. Forty-five mathematicians attended and the opening address, on 'The present state of mathematics', was given by Felix Klein of Göttingen, one of just four participants from outside the USA.

The first official Congress was held in Zürich in 1897, where it was decided to hold such international meetings every three to five years, and it was at the next one, in Paris in 1900, that Hilbert presented his famous lecture on the future problems of mathematics. Since these early gatherings, more than a score of international congresses have been held around the world, usually every four years.

These meetings usually take place without incident, but there have been a few difficulties along the way. The 1920 and 1924 Congresses were boycotted by many mathematicians, since Germans and Austrians were excluded, while the 1982 Warsaw Congress had to be postponed for a year due to the introduction of martial law in Poland.

Two medallists have been unable to attend due to visa restrictions, while another has declined the award.

THE FIELDS MEDAL

John Charles Fields was a mathematics professor at the University of Toronto, and President of the Toronto Congress in 1924. The profit from this meeting, together with later money from his estate, provided funding for the 'International Medals for Outstanding Discoveries in Mathematics', now known as Fields Medals. First awarded in 1936, the gold medals are produced by the Royal Canadian Mint and feature Archimedes on one side and an inscription on the other.

INTERNATIONAL CONGRESSES AND FIELDS MEDAL WINNERS

In the table we show the country with which each medallist is mainly associated.

Congress	Fields Medal winners
1897: Zürich, Switzerland	–
1900: Paris, France	–
1904: Heidelberg, Germany	–
1908: Rome, Italy	–
1912: Cambridge, UK	–
1920: Strasbourg, Germany	–
1924: Toronto, Canada	–
1928: Bologna, Italy	–
1932: Zürich, Switzerland	–
1936: Oslo, Norway	Lars Ahlfors (Finland); Jesse Douglas (USA)
1950: Cambridge, USA	Laurent Schwartz (France), Atle Selberg (Norway)
1954: Amsterdam, Netherlands	Kunihiko Kodaira (Japan/USA), Jean-Pierre Serre (France)
1958: Edinburgh, UK	Klaus Roth (UK), René Thom (France)
1962: Stockholm, Sweden	Lars Hörmander (Sweden), John Milnor (USA)
1966: Moscow, USSR	Michael Atiyah (UK), Paul Cohen and Stephen Smale (USA), Alexander Grothendieck (Germany)
1970: Nice, France	Alan Baker (UK), Heisume Hironaka (Japan), Sergei Novikov (USSR), John Thompson (USA)
1974: Vancouver, Canada	Enrico Bombieri (Italy), David Mumford (USA)
1978: Helsinki, Finland	Pierre Deligne (Belgium), Grigory Margulis (USSR), Charles Fefferman and Daniel Quillen (USA)
1983: Warsaw, Poland	Alain Connes (France), William Thurston (USA), Shing-Tung Yau (China)
1986: Berkeley, USA	Simon Donaldson (UK), Gerd Faltings (Germany), Michael Freedman (USA)
1990: Kyoto, Japan	Vladimir Drinfel'd (USSR), Shigefumi Mori (Japan), Vaughan Jones (New Zealand), Edward Witten (USA)
1994: Zürich, Switzerland	Jean Bourgain (Belgium), Pierre-Louis Lions and Jean-Christophe Yoccoz (France), Efim Zelmanov (Russia)
1998: Berlin, Germany	Richard Borcherds and Timothy Gowers (UK), Maxim Kontsevich (France/Russia), Curtis McMullen (USA)
2002: Beijing, China	Laurent Lafforgue (France), Vladimir Voevodsky (Russia)
2006: Madrid, Spain	Andrei Okounkov and Grigori Perelman (Russia), Terence Tao (Australia/USA), Wendelin Werner (France)
2010: Hyderabad, India	Elon Lindenstrauss (Israel), Ngô Bao Châu (Vietnam), Stanislav Smirnov (Russia), Cédric Villani (France)

ABEL PRIZEWINNERS

In June 2002, to commemorate the bicentenary of Abel's birth, the Norwegian Academy of Science and Letters launched the **Abel Prize,** to be presented annually by the King of Norway for outstanding scientific work in the field of mathematics.

2003: Jean-Pierre Serre (France)
2004: Michael Atiyah (UK) and Isadore Singer (USA)
2005: Peter Lax (Hungary/USA)
2006: Lennart Carleson (Sweden)
2007: Srinivasa Varadhan (India/USA)
2008: John Thompson (USA) and Jacques Tits (France)
2009: Mikhail Gromov (Russia)
2010: John Tate (USA)
2011: John Milnor (USA)
2012: Endre Szemerédi (Hungary)

FURTHER READING

Here is a selection of sources that the authors have found interesting and useful.

Books

General interest books on mathematics and its history

D. Acheson, *1089 and all that,* Oxford, 2010.

M. Anderson, V. Katz and R. Wilson, *Sherlock Holmes in Babylon, and Other Tales of Mathematical History,* Mathematical Association of America, 2004.

M. Anderson, V. Katz and R. Wilson, *Who Gave You The Epsilon?, and Other Tales of Mathematical History,* Mathematical Association of America, 2004.

E. Behrends, *Five-Minute Mathematics,* American Mathematical Society, 2008.

W. P. Berlinghoff and F. Q. Gouvêa, *Math through the Ages: A Gentle History for Teachers and Others,* Oxton House and the Mathematical Association of America, 2004.

R. Courant and H. Robbins, *What is Mathematics?,* Oxford, 1996.

T. Crilly, *50 Mathematical Ideas You Really Need to Know,* Quercus, 2008.

K. Devlin, *Mathematics: The New Golden Age,* Columbia, 1999.

K. Devlin, *The Millennium Problems: The Seven Greatest Unsolved Mathematical Puzzles of Our Time,* Basic Books, 2003.

J. Farndon, *The Great Scientists,* Arcturus, 2005.

L. Hodgkin, *A History of Mathematics: Mesopotamia to Modernity,* Oxford, 2005.

S. Hollingdale, *Makers of Mathematics,* Penguin, 1989.

C. A. Pickover, *The Math Book: From Pythagoras to the 57th Dimension, 250 Milestones in the History of Mathematics,* Sterling, 2009.

A. Rooney, *The Story of Mathematics,* Arcturus, 2009.

G. Simmons, *Calculus Gems: Brief Lives and Memorable Mathematics,* McGraw-Hill, 1992.

I. Stewart, *From Here to Infinity,* Oxford, 1996.

I. Stewart, *Taming the Infinite,* Quercus, 2008.

S. M. Stigler, *The History of Statistics,* Harvard, 1986.

History of mathematics texts

C. B. Boyer and U. C. Merzbach, *A History of Mathematics,* Wiley, 1991.

D. Burton, *The History of Mathematics: An Introduction,* McGraw-Hill, 2010.

H. Eves, *An Introduction to the History of Mathematics,* Thomson Brooks/Cole, 1990.

I. Grattan-Guinness, *The Fontana History of the Mathematical Sciences,* Fontana, 1997.

V. J. Katz, *A History of Mathematics,* Pearson, 2008.

M. Kline, *Mathematical Thought from Ancient to Modern Times,* Oxford, 1990.

J. Stillwell, *Mathematics and its History,* Springer, 2010.

D. Struik, *A Concise History of Mathematics,* Dover, 1987.

Source books containing work translated where necessary into English

L. Berggren, J. Borwein and P. Borwein, *Pi: A Source Book,* Springer, 1999.

R. Calinger, *Classics of Mathematics,* Prentice Hall, 1999.

J. Fauvel and J. Gray, *The History of Mathematics – a Reader,* Mathematical Association of America, 1997.

V. J. Katz, *The Mathematics of Egypt, Mesopotamia, China, India, and Islam: A Sourcebook,* Princeton, 2007.

D. Smith, *A Source Book in Mathematics,* Dover, 1959.

J. Stedall, *Mathematics Emerging: A Sourcebook 1540–1900,* Oxford, 2008.

More specialized topics

A. D. Aczel, *The Artist and the Mathematician: The Story of Nicolas Bourbaki, The Genius Mathematician who Never Existed*, High Stakes, 2007.

B. Artmann, *Euclid: The Creation of Mathematics*, Springer, 1999.

L. Berggren, *Episodes in the Mathematics of Ancient Islam*, Springer, 2003.

B. Collier and J. MacLachlan, *Charles Babbage and the Engines of Perfection*, Oxford, 1998.

J. Fauvel, R. Flood and R. Wilson (eds.), *Oxford Figures: 800 Years of the Mathematical Sciences*, Oxford, 2000.

R. Flood, A. Rice, and R. Wilson (eds.), *Mathematics in Victorian Britain*, Oxford, 2011.

J. V. Field, *The Invention of Infinity: Mathematics and Art in the Renaissance*, Oxford, 1997.

D. H. Fowler, *The Mathematics of Plato's Academy*, Oxford, 1998.

J. Gray, *Worlds Out of Nothing: A Course in the History of Geometry in the 19th Century*, Springer, 2007.

J. J. Gray, *János Bolyai, Non-Euclidean Geometry and the Nature of Space*, MIT Press, 2004.

G. G. Joseph, *The Crest of the Peacock; The Non-European Roots of Mathematics*, Princeton, 2011.

W. R. Knorr, *The Ancient Tradition of Geometric Problems*, Dover, 1986.

O. Ore, *Number Theory and its History*, Dover, 1988.

D. O'Shea, *The Poincaré Conjecture*, Walker and Co., 2007.

P. Pesic, *Abel's proof, An Essay on the Sources and meaning of Mathematical Unsolvability*, M.I.T. Press, 2003.

K. Plofker, *Mathematics in India*, Princeton, 2008.

E. Robson, *Mathematics in Ancient Iraq: A Social History*, Princeton, 2008.

S. Singh, *Fermat's Last Theorem*, Fourth Estate, 1997.

J. Stedall, *A Discourse Concerning Algebra: English Algebra to 1685*, Oxford, 2002.

T. Körner, *The Pleasures of Counting*, Cambridge, 1996.

G. Van Brummelen, *The Mathematics of the Heavens and the Earth: The Early History of Trigonometry*, Princeton, 2009.

R. Wilson, *Four Colours Suffice: How the Map Problem was Solved*, Penguin, 2003.

R. Wilson, *Lewis Carroll in Numberland: His Fantastical Mathematical Logical Life*, Penguin, 2009.

B. Yandell, *The Honors Class: Hilbert's Problems and Their Solvers*, A. K. Peters, 2002.

Biographies

W. K. Bühler, Gauss: A Biographical Study, Springer, 1987.

T. Crilly, *Arthur Cayley: Mathematician Laureate of the Victorian Age*, Johns Hopkins University Press, 2006.

W. Dunham, *The Genius of Euler: Reflections on his Life and Work*, Mathematical Association of America, 2007.

R. Flood, M. McCartney and A. Whitaker (eds.), *Kelvin, Life, Labours and Legacy*, Oxford, 2008.

R. Kanigel, *The Man Who Knew Infinity: A Life of the Genius Ramanujan*, Washington Square Press, 1992.

K. Parshall, *James Joseph Sylvester; Life and Work in Letters*, Oxford, 1998.

C. Reid, *Hilbert*, Springer, 1996.

C. Reid, *Julia* [Robinson], *A Life in Mathematics*, Mathematical Association of America, 1996.

S. Stein, *Archimedes: What Did He Do Besides Cry Eureka?*, Mathematical Association of America, 1999.

B. A. Toole, *Ada, The Enchantress of Numbers: A Selection from the Letters of Lord Byron's Daughter and Her description of the First*

Computer, Critical Connection, 1998.
R. S. Westfall, *Never at Rest: A Biography of Isaac Newton,* Cambridge, 1980.

General reference books

F. Cajori, *A History of Mathematical Notations,* Dover, 1993.
T. Gowers (ed.), *The Princeton Companion to Mathematics,* Princeton, 2008.
D. Nelson, *The Penguin Dictionary of Mathematics,* Penguin, 2008.
E. Robson and J. Stedall (eds.), *The Oxford Handbook of the History of Mathematics,* Oxford, 2009.

Internet

The Digital Mathematics Library gives access to many original works: *www.mathematik.uni-bielefeld.de/~rehmann/DML/dml_links.html*

The MacTutor mathematical archive is a useful source of biographical data and other articles on various themes: *www-history.mcs.st-and.ac.uk/*

Wolfram Mathworld has information on various mathematical topics: *mathworld.wolfram.com/*

Information about the British Society for the History of Mathematics: *www.dcs.warwick.ac.uk/bshm/*

Information about history of mathematics for young mathematicians: *www.mathsisgoodforyou.com/*

Information about an Open University introductory historical course, *TM190, The Story of Maths*, based on Marcus du Sautoy's BBC television series:

www3.open.ac.uk/study/undergraduate/course/tm190.htm

Information about an Open University M.Sc. dissertation course, *M840, The History of Modern Geometry Maths*: *www3.open.ac.uk/study/postgraduate/course/m840.htm*

Information about a course based on Marcus du Sautoy's *Number Mysteries* (Fourth Estate, 2011) at the Department of Continuing Education, Oxford University: *www.conted.ox.ac.uk*

PICTURE CREDITS

Corbis: 9, 20, 24, 32, 33(l), 39, 49, 54, 56(t), 72, 78(b), 96, 103(t), 108, 114(b), 139, 141, 144, 154(l), 162, 164, 166, 174, 176, 184, 185, 197, 199

Science Photo Library: 37, 46(t), 62(t), 65(t), 74, 87, 88(t), 142(l), 156(t), 172, 182, 194

Getty: 102(t), 110, 125

Topfoto: 38, 41, 105(b), 116, 134(x2), 150(b), 154(r), 181, 187

British Museum: 57

RIA Novosti: 42

Mary Evans: 60

Bridgeman: 66(t), 84, 119, 126, 189

(Courtesy of) The American Mathematical Society: 190

Bill Stoneham: 145

Science and Society: 149

Trinity College, Cambridge Archive: 180

Biblioteca Ambrosiana: 65(b)

Clipart: 27, 28, 33, 150(t)

Shutterstock: 29, 73(b), 123, 129(t)

Diagrams and maps: David Woodroffe

We have endeavoured to contact the copyright-holders of all images used in this book. Any oversights in this regard will be rectified in future editions

INDEX

Names in **bold type** refer to the mathematicians especially featured in this book.